Photo Pantanal

Wade!
Wish you the best—
Tom J. Ulrich

Photo Pantanal

Tom J. Ulrich

AMERICAN WILDLIFE
WEST GLACIER, MT 59936
2002

Library of Congress Cataloging-in-Publication Data

Ulrich, Tom J.
Photo Pantanal / Tom Ulrich.
p. cm.
ISBN 0-9715264-0-0 (alk. paper)
1. Zoology—Pantanal—Pictorial works. 2. Pantanal—Pictorial works. I. Title.

QL242.U46 2002
591.981—dc21
2002021446

PRINTED IN HONG KONG

AMERICAN WILDLIFE
P.O. Box 361 • West Glacier, MT 59936
406-387-5521

Dedicated to

William S. Sheldon

&

Rubens Cruz Pereira

Acknowledgements

Many individuals were instrumental in the formation and fine-tuning of this pictorial feature into the final product now held in your hands, but three were important indirectly. They are my best friend Bill Sheldon, and Brazilians Rubens Cruz Pereira and Jovanil De Camargo. I had never heard of the Pantanal until Bill Sheldon called my attention to the area. His daughter was a foreign exchange student in Cuiaba, Mato Grosso residing with the Pereira family. Upon a visit to Cuiaba, Bill made a side trip to the Pantanal. His only statement to me was, "You have to see this place!" The family hosting Bill's daughter was Rubens Cruz Pereira. Rubens became my contact over the many trips needed to capture the splendor of the Pantanal on film. Rubens, beside being a super friend, when time permitted acted as my guide, but mostly was there to find great deals on vehicle rentals and make reservations. Eventually he secured Jovanil De Camargo to drive and wait patiently for me to photograph. To these three individuals I am in deep gratitude.

When doing a book such as this, untold questions are a certainty. Mountain Press in Missoula, MT was there to answer them. Two Alaskan friends, Chris Batin and Adela Batin Jackson offered their years of wisdon in publishing with much positive advice and ideas. Julia Brewer, Gail Jokerst and Bert Gildart edited my manuscript for grammar and punctuation. Caroline Orlandi translated my English copy into a Portuguese version. To all of these individuals I owe a special appreciation.

Finally, a special thank you goes to Eastern Illinois University's Council on Faculty Research for awarding a Summer Research Grant to Deborah A. Woodley, a professor in the School of Technology. The grant money provided funding for her involvement in the design and editing of the book, publication software and the translation of the text from English to Portuguese.

Introduction

Brazilians as well as visitors around the world are starting to discover the Pantanal. They come to experience the solitude and vastness the area presents, but mostly to observe the land's wide variety of wildlife. Fascinated by the diversity and density of the fauna I, too, have been lured to the area. Since 1990 I have made eight excursions to photograph wildlife. During each of my visits the animals were readily visible and, most of the time, they seemed unconcerned by human presence.

The purpose of PHOTO PANTANAL is to share visual images of the Pantanal with those who might never get a chance to experience the area personally, and to document the area's beauty and its life on film before change can occur. As well, the book might help first-time visitors identify many of the area's wildlife species. Hopefully, the book will also recall a glorous adventure as visitors re-turn the pages in future years.

You can see most of the birds featured in PHOTO PANTANAL during a three- or four-day visit. However, the wildlife here is greatly influenced by the wet/dry seasonal cycle. The perpetual rains of the wet season flood the area and disperse the fish and wildlife that feed upon these fish. Accessibility to the region is also a problem from November to April as roads turn to gumbo. Then, even the locals find travel difficult. The best time to visit is at the end of the dry season, July to September, when fauna congregates at the shrinking water holes full of stranded fish. At this time birds, reptiles and some mammals feast on fish, fulfilling their role in the area's timeless and dramatic interplay of predator and prey.

PHOTO PANTANAL calls attention to how special this area really is. My hope is that it will raise the level of awareness for this priceless area, helping, in some way, to preserve for future generations the Pantanal's unparalleded beauty.

Table of Contents

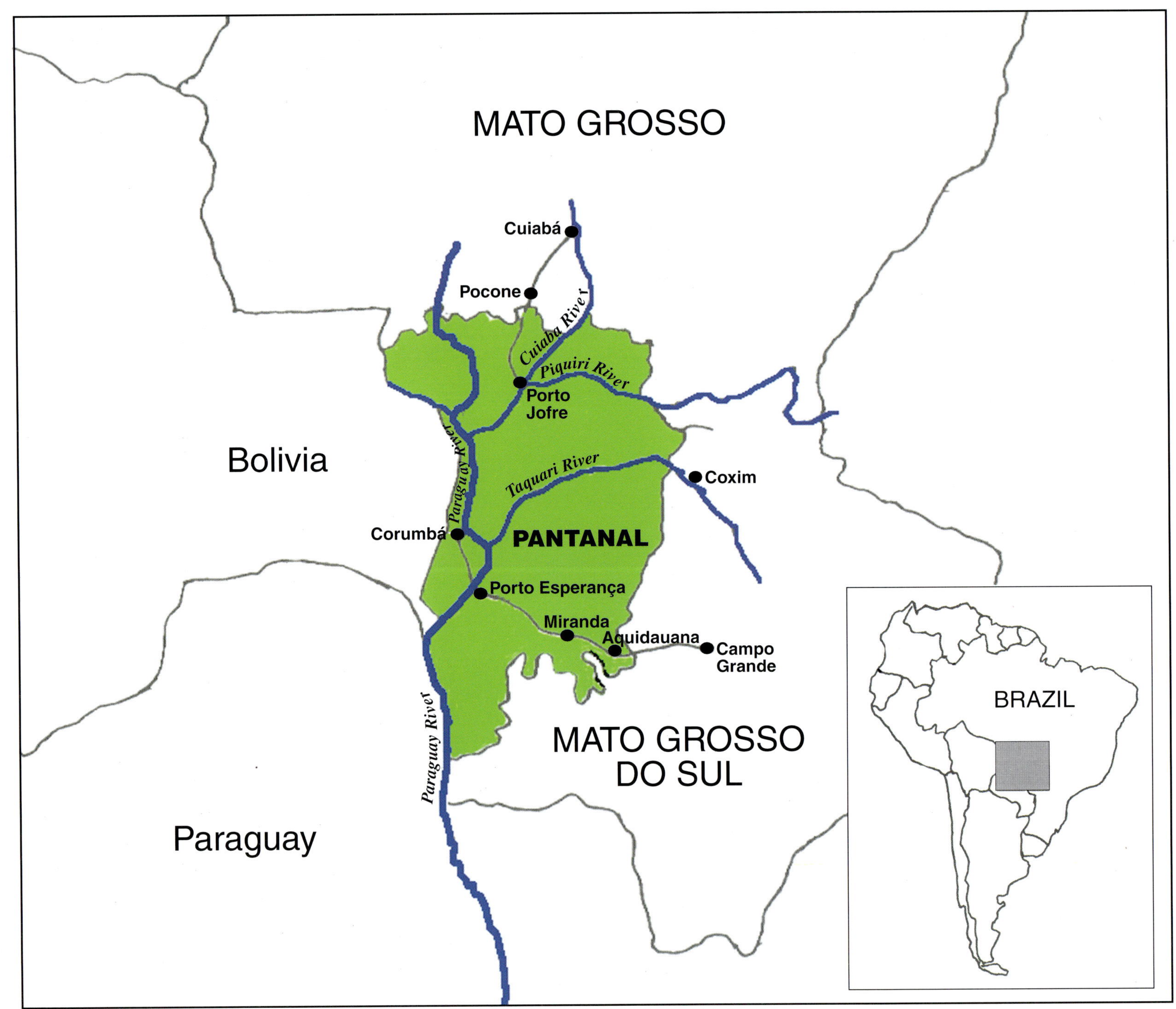

Map of Pantanal

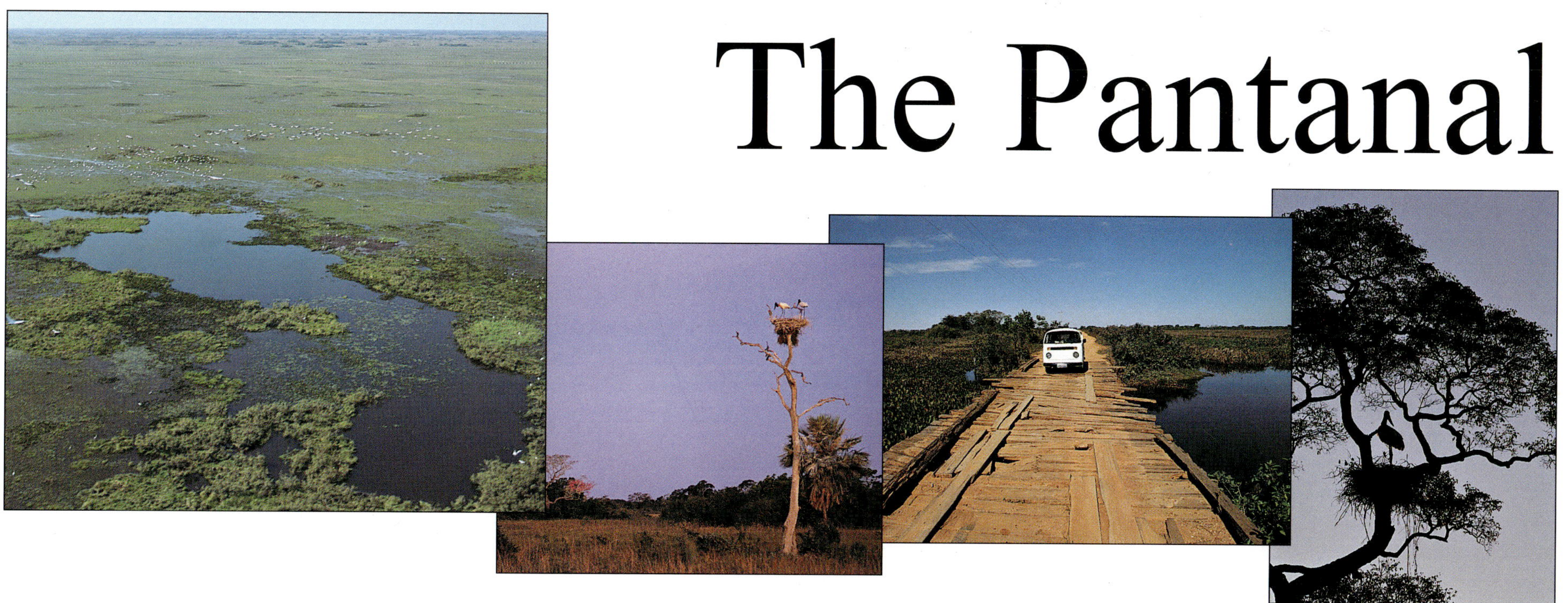

The Pantanal

Largely located in southwest Brazil, the Pantanal contains the upper portion of the Paraguay River and its tributaries. Actually, it comprises about one third of the Upper Paraguay River Basin. About 70 percent of the land area for the Pantanal is located in Brazil; the rest of it is spread between Bolivia and Paraguay. In Brazil it is divided between the two states of Mato Grosso and Mato Grosso Do Sul with the latter having the bigger share. Best estimates put the entire Pantanal at 68,000 square miles. It is believed this region was once connected to the ocean, forming an inland sea. The marine fossils found buried beneath layers of sediment which over time have been gradually filling the basin confirm this theory.

Pantanal, derived from the word pantano, means marsh or swamp. It designates a huge flood plain, what geologists call an immense alluvial depression. The area averages a drop of one inch in elevation for every mile in the north/south gradient. In the east/west gradient the drop in elevation is a little over three inches every mile. During the rainy season from October to April, the average annual precipitation of 32 inches causes the rivers to overflow, flooding 75 percent of the Pantanal. Fish swim out over the flooded plain where most spawn. With the receding waters of the dry season untold numbers of fish are left behind in shallow pools. They become a rich food source for the predators located there.

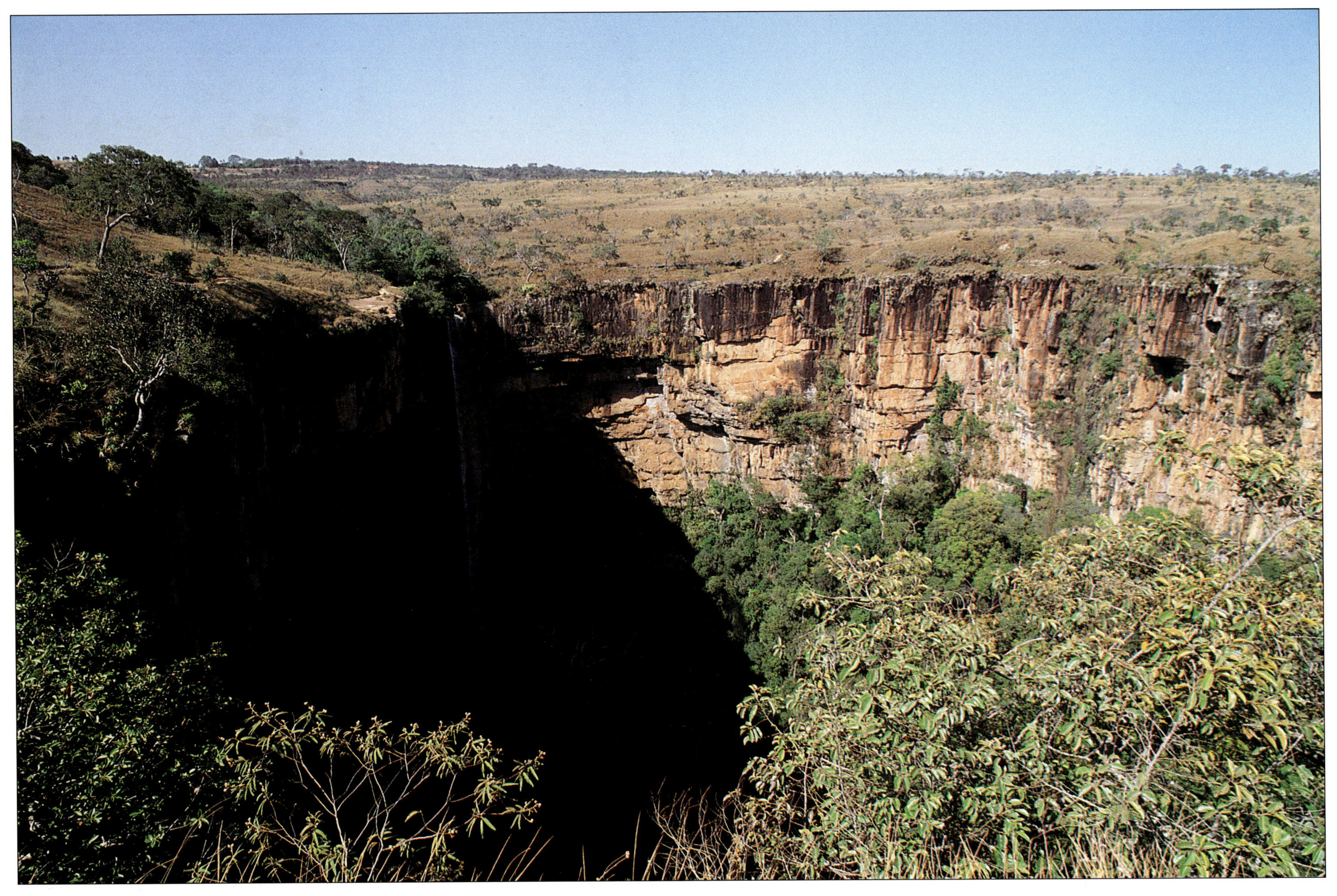

Chapada

As mentioned earlier, the Pantanal comprises about one third of the Upper Paraguay River Basin. The remaining two thirds is known as the "Planalto" or highlands. The upper end of this highlands is a slightly pronounced rise actually forming a physical barrier between the Upper Paraguay River Basin and the Amazon. Sediments carried from these highlands have filled in the Pantanal over eons, creating the immense alluvial fan.

In Cuiaba, the locals call the nearby highlands Chapada dos Guimaraes or just simply "Chapada." Numerous spectacular waterfalls created from drainage of the highlands to the flood plain run clear and cool. In the dry season this area becomes an important recreational get-a-way for the locals.

Rainfall in the highlands is much greater than in the Pantanal and can average 64 inches a year with most of that falling in the rainy season. This vast amount of water is extremely important to the area's hydraulic cycle. It provides the needed second punch that keeps flushing the Pantanal with fresh water.

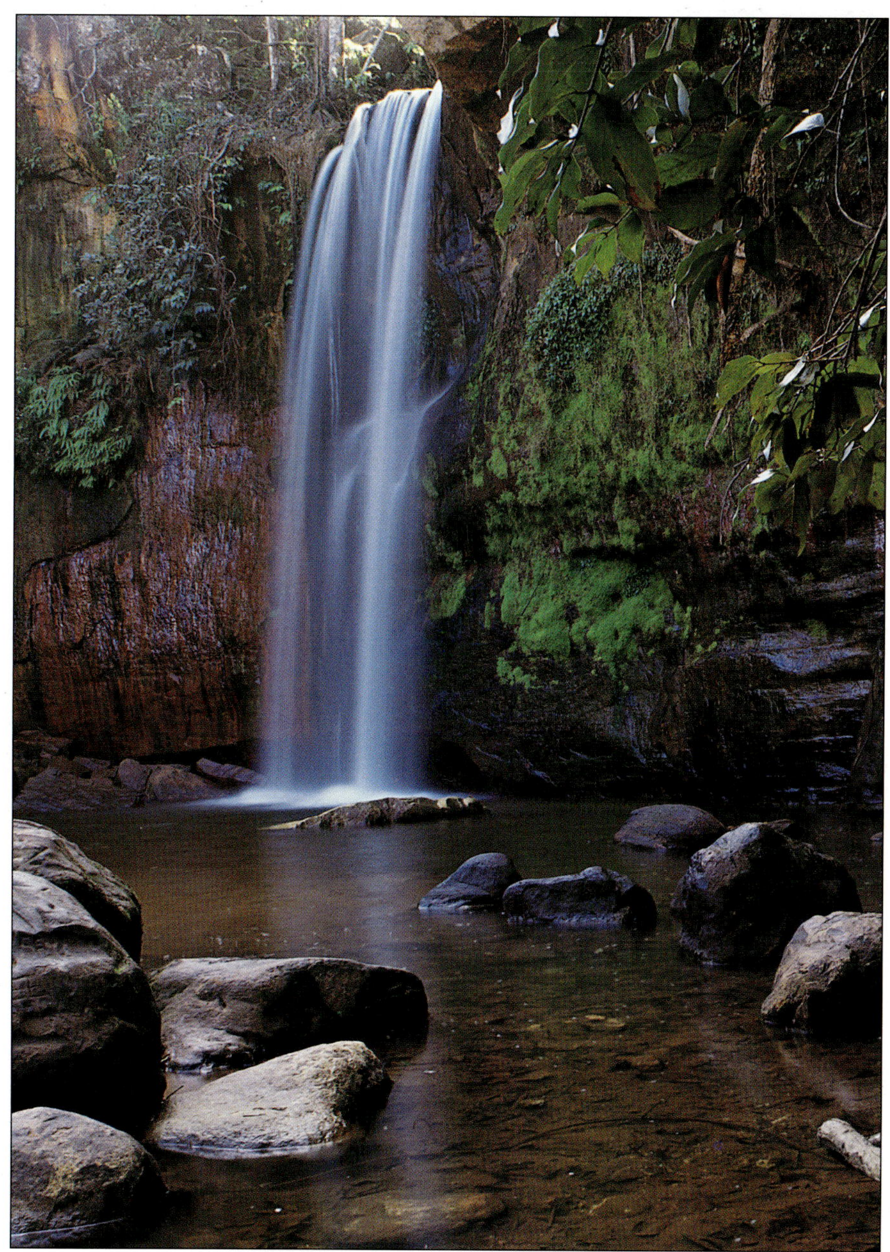

Waterfall from the higher basin which eventually flows to the Pantanal.

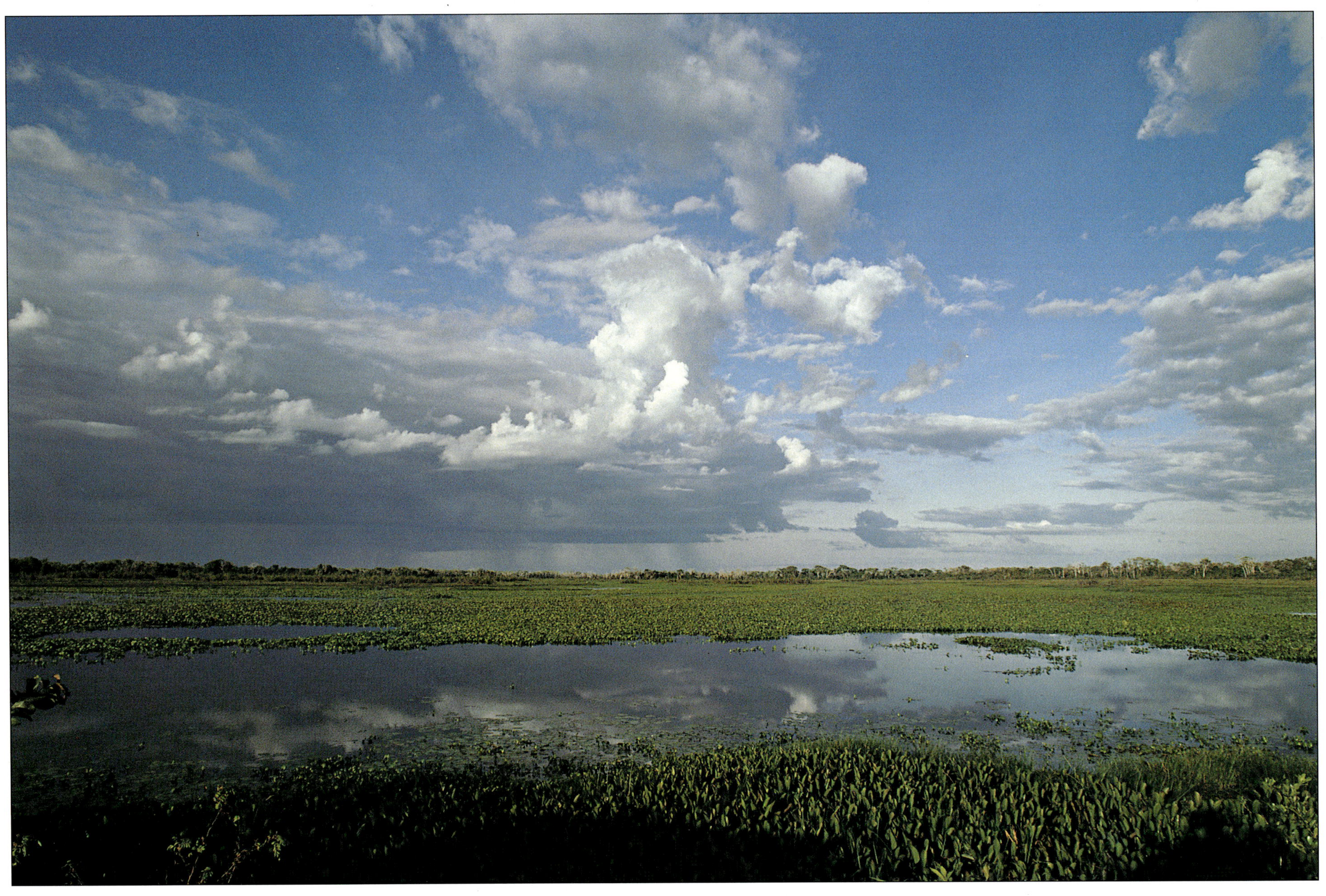

Vastness of the Pantanal

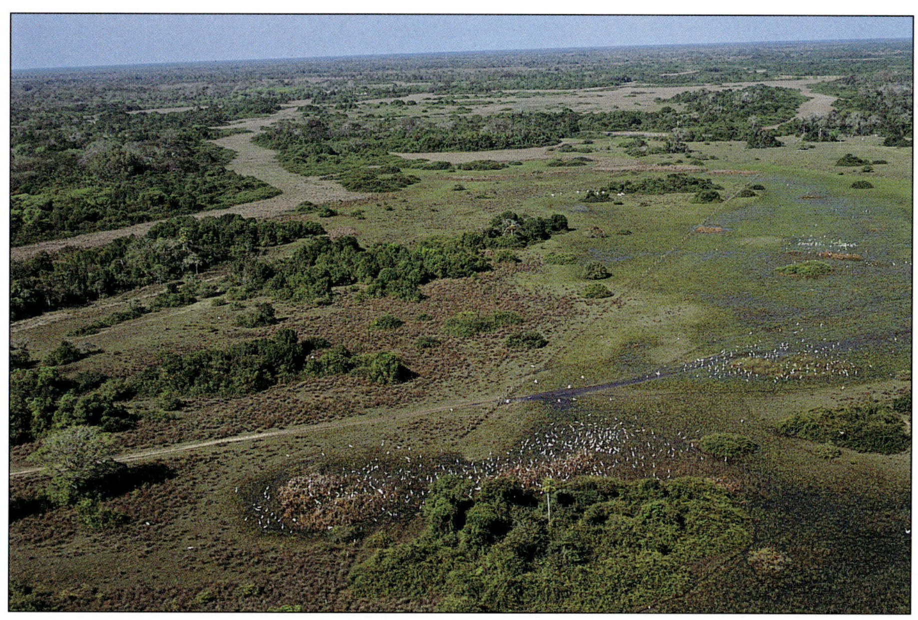

Ever see anything flatter?

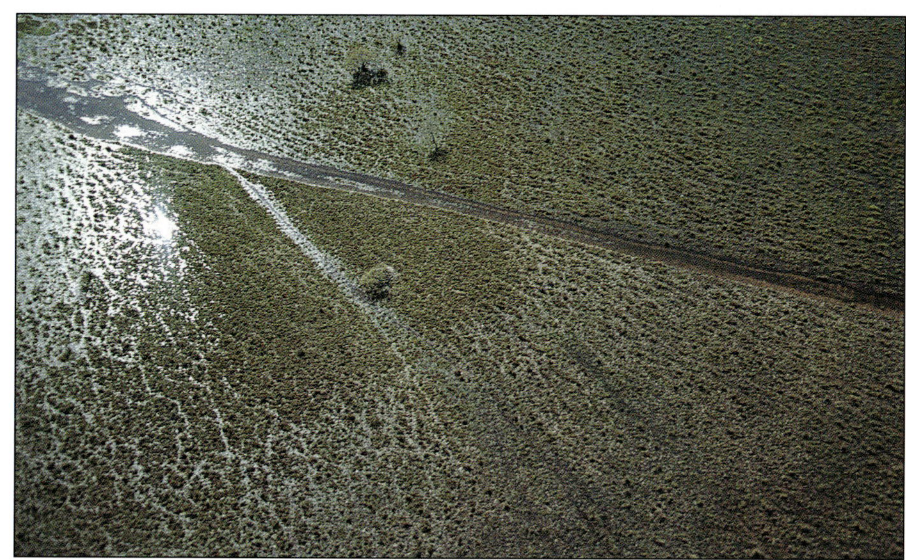

This is an excellent aeriel photo showing how wet the land can be.

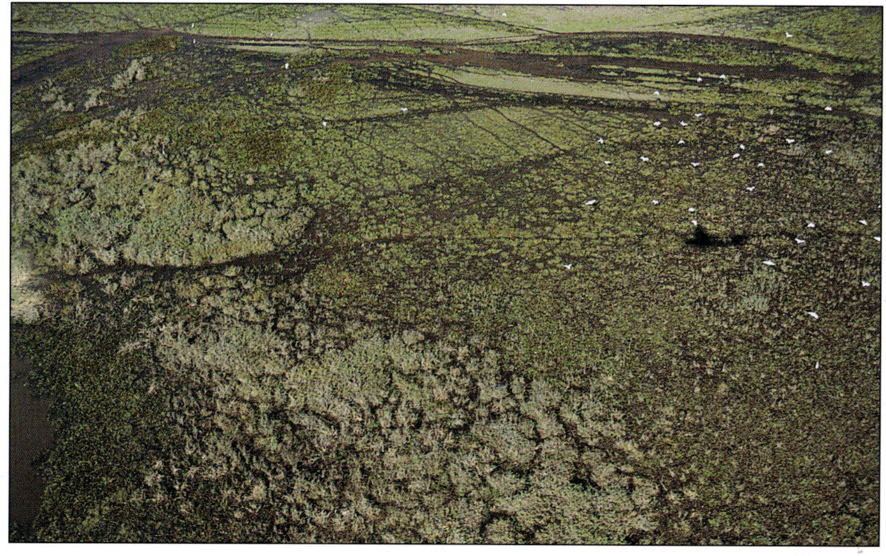

An aeriel photo showing some of the bird life.

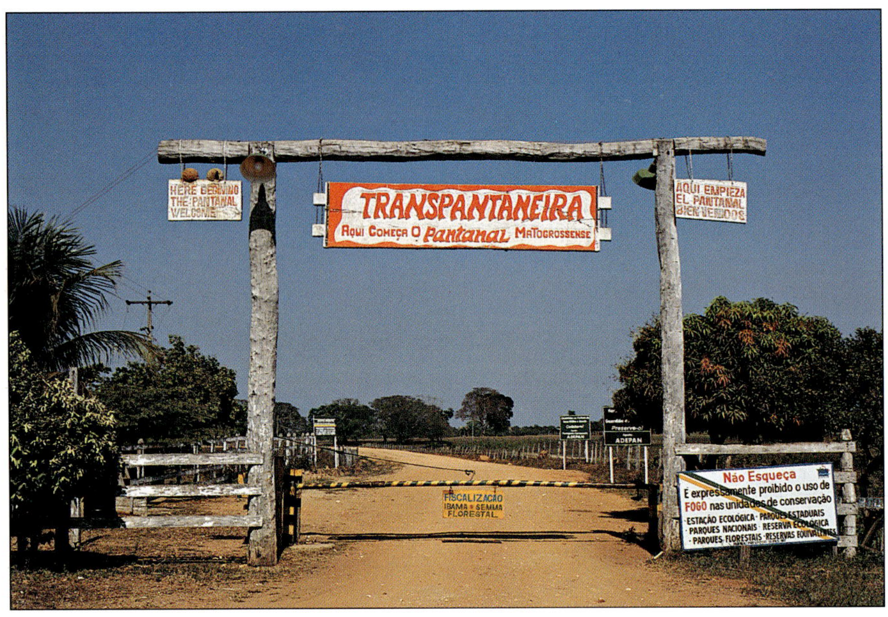

Entrance to the Transpantaneria
highway at mile 10.

In 1971, work began to construct a road across the Pantanal. It would start at Pocane and finish at Corumba, a distance of 211 miles and would be called the "Transpantaneria." However, all work stopped 100 miles into the project at Porto Jofre on the Cuiaba River. To construct the road, earth was excavated in two ditches and deposited between them, forming the roadbed. The canals left by the excavation are continuously filled with water, producing excellent fish habitat and attracting all the predators that feed upon them. Due to the large volume of water moving about, 125 wooden bridges had to be built. Many have deterioated, but currently a concerted effort is being made to rebuild the bridges, starting at the beginning. At mile 10, a wooden gate officially denotes the beginning of the Pantanal. Visitors pay a small fee to enter.

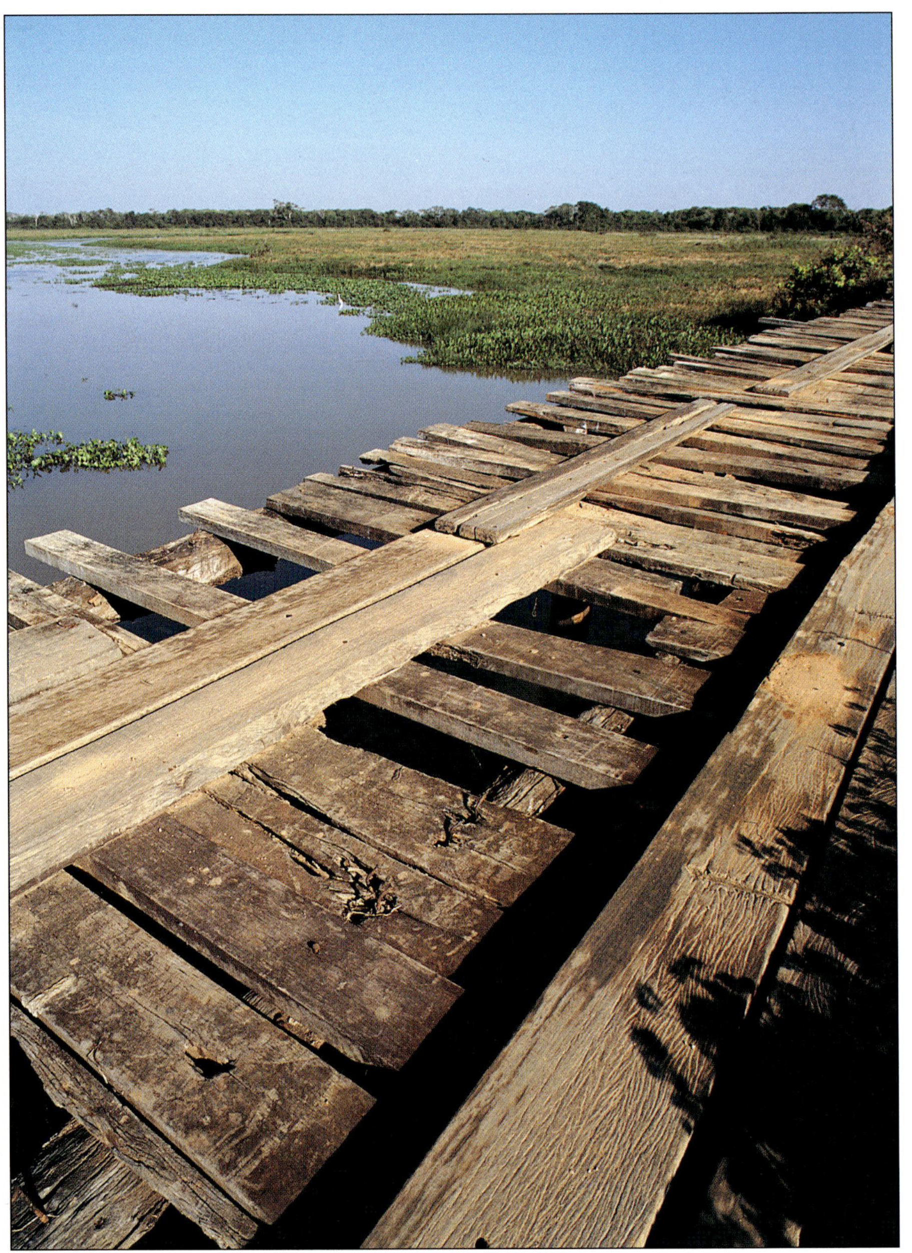

One of the many bridges found along
the Transpantaneria highway.

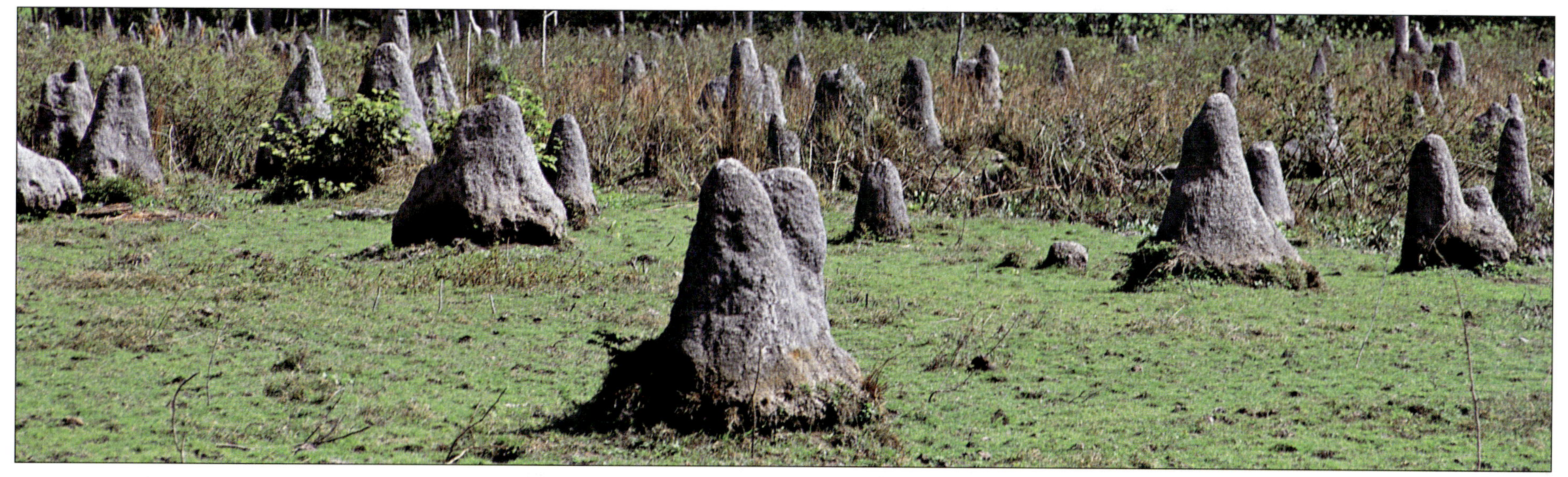

Termite mounds

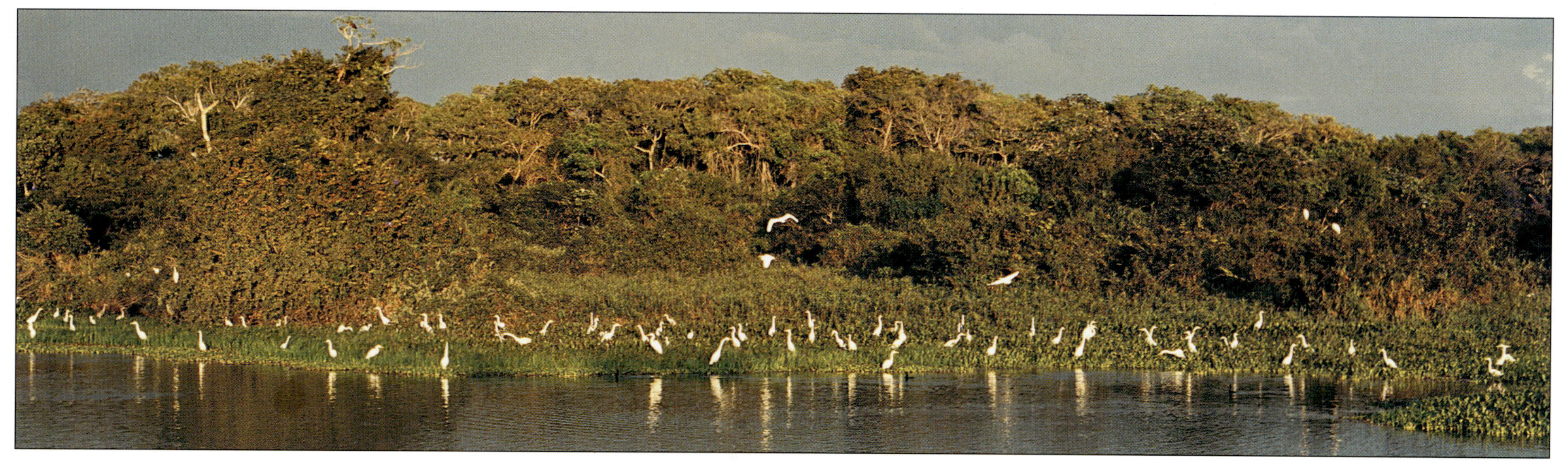

Some of the birds seen from the road

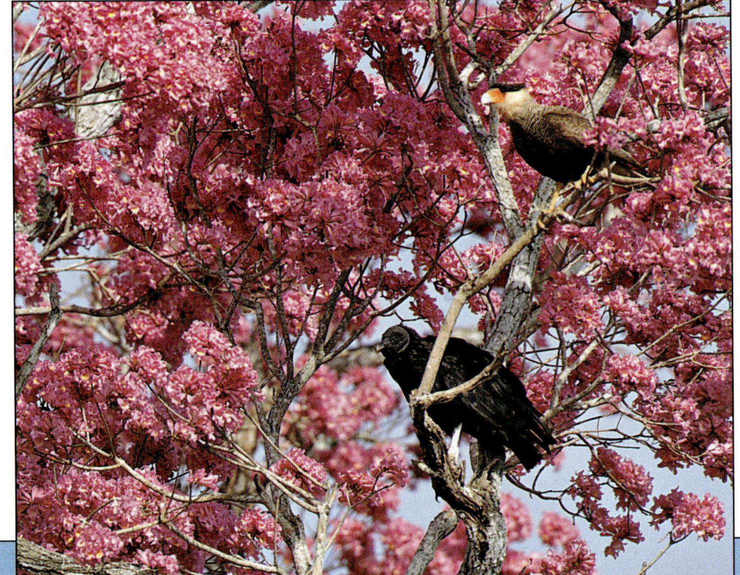

On a visit during the dry season one will observe occasional trees brightly colored in various shades of purple. These are piuva trees in full bloom. They dot the landscape and definitely add to any attempt in scenic photography. Also common along any shallow bodies of water are pond lilies in bloom. These showy flowers are likewise found in various shades of purple. One rare shallow water species seen are giant lily pads, which can grow to five feet across. It is a special experience to come across this plant.

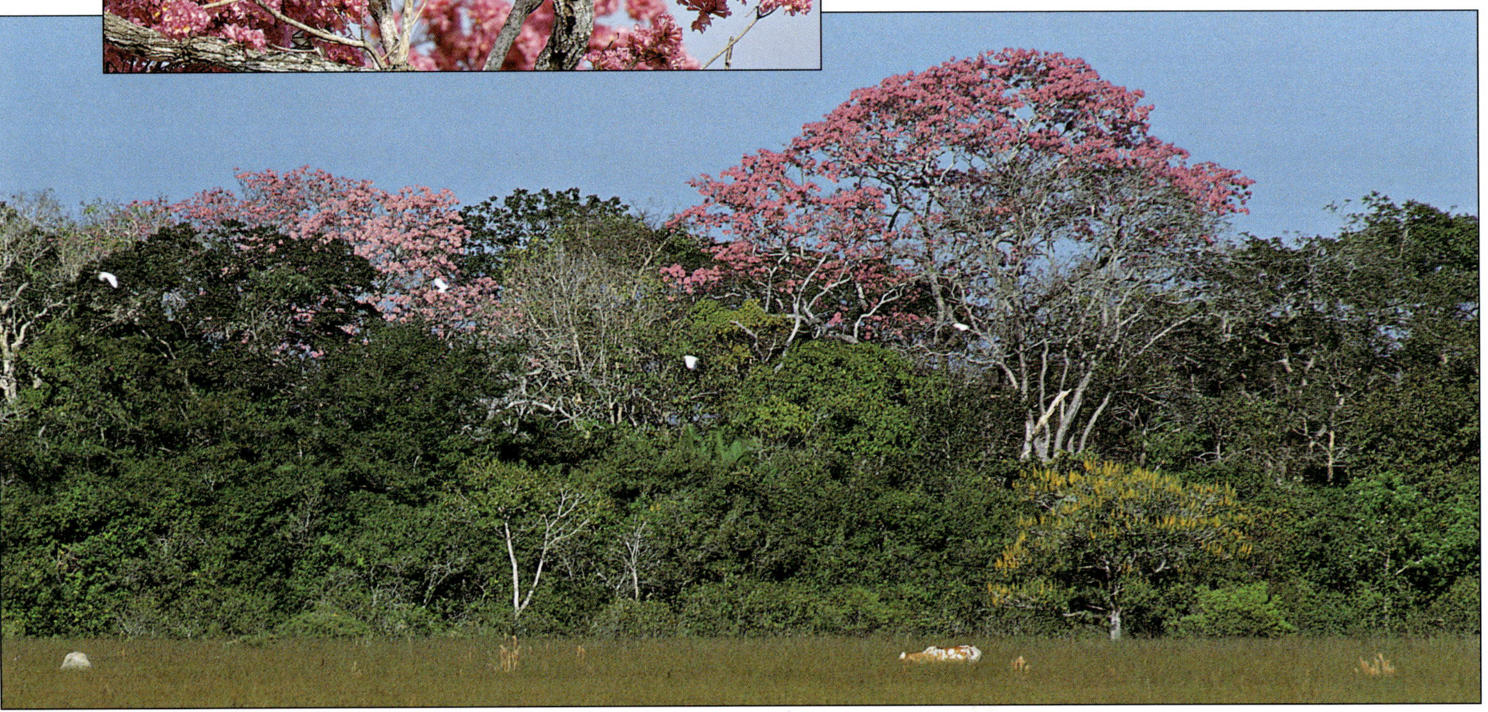

Piuva Trees in bloom
Tecoma sp.

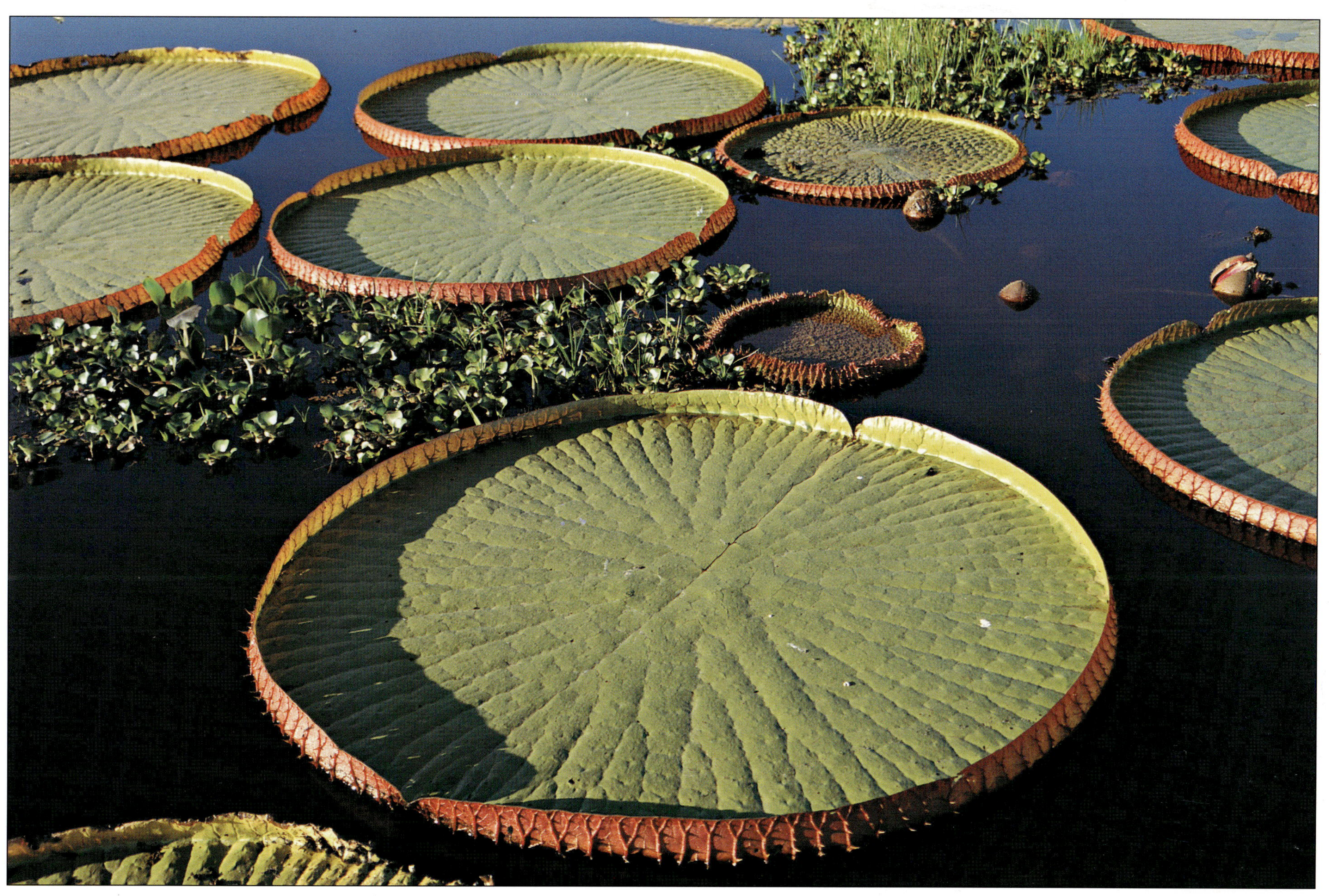

Giant Lilypads
Victoria cruziana

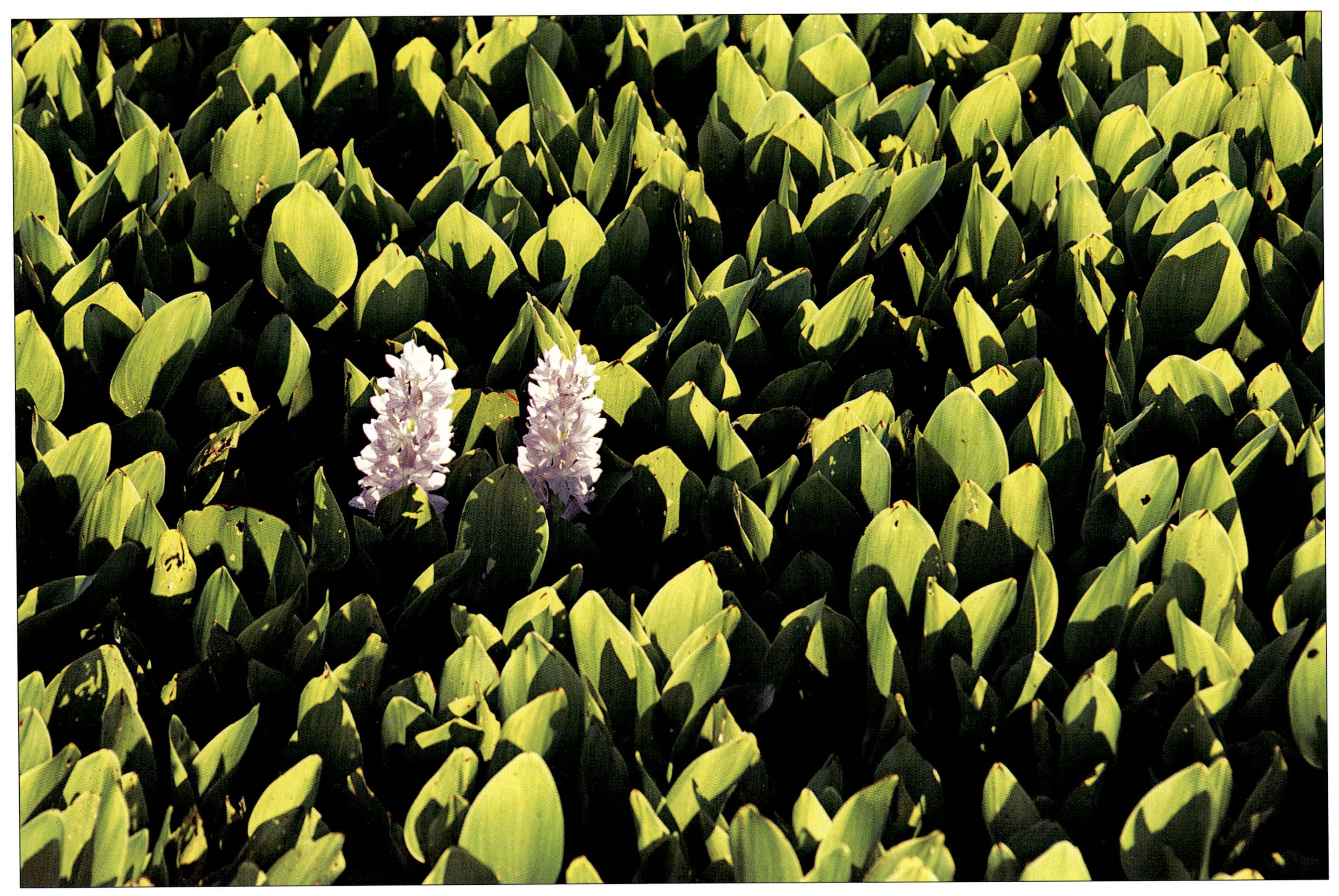

Pond Lilies

Eichhornia sp

The People

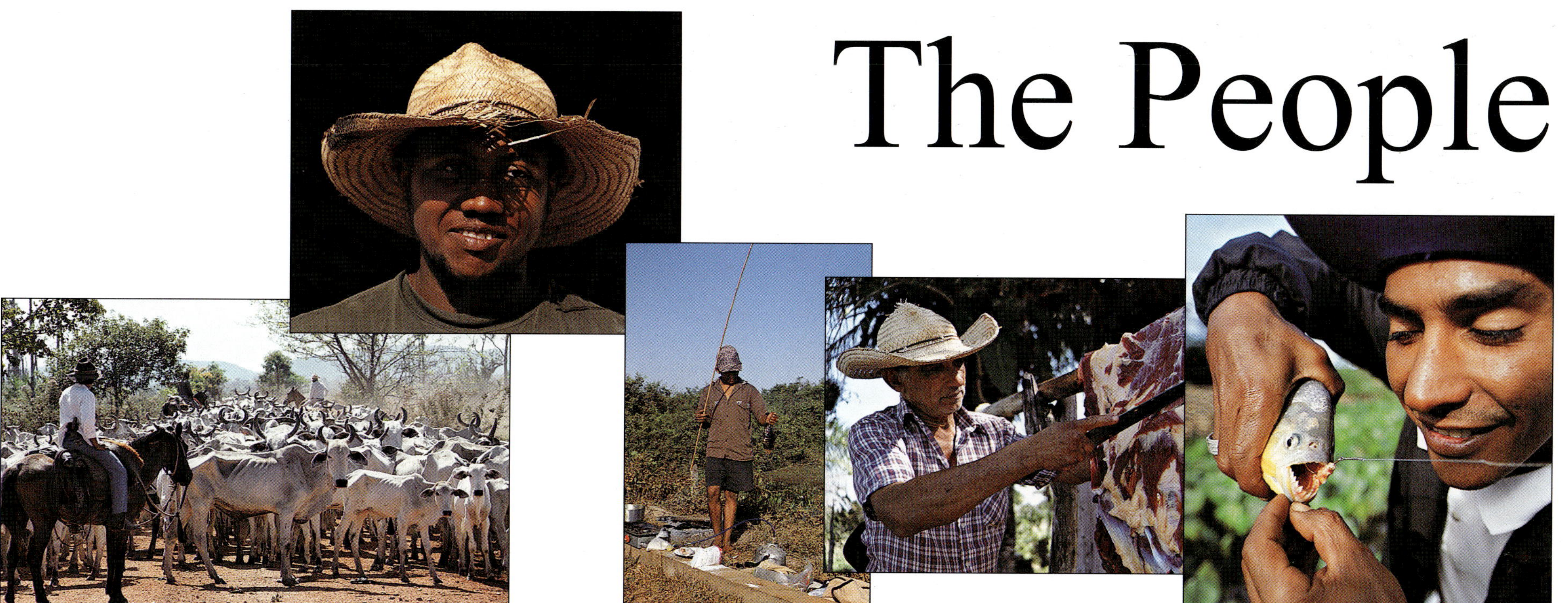

The first indigenous inhabitants of the Pantanal were Indians of several tribes, including the Guaianis, Bororos, Paiaguas and Guatos. They had free range of the land until the early 1800's when colonization began. Some of the tribes mixed with the colonizers, but most remained separate. The African slaves that came by way of San Paulo and the Paraguays were most successful in intermingling with the natives, giving rise to the actual Pantaneiro.

While the quest for gold was one of the primary reasons for the immigrant migration, it did not last long as the gold played out. Most immigrants then turned to cattle ranching even though the large expanse of land was just barely suitable for grazing. Several kinds of European cattle were introduced. From the interbreeding of these species the Pantaneiro bull appeared. It differs from the Indian brahma bull by the lack of the hump on its back. Like the humans that adapted to the environment, a horse was bred that was perfectly suited to the marsh conditions found in and around the ranches of the region. The horse is also called Pantaneiro.

With the long distances and difficult access, the pantaneiro man must get accustomed to the isolation and loneliness. Along with their diet of meat, fish, beans, rice, manioc and manioc flour, they also learned to depend on wild fruits, roots and leguminous plants.

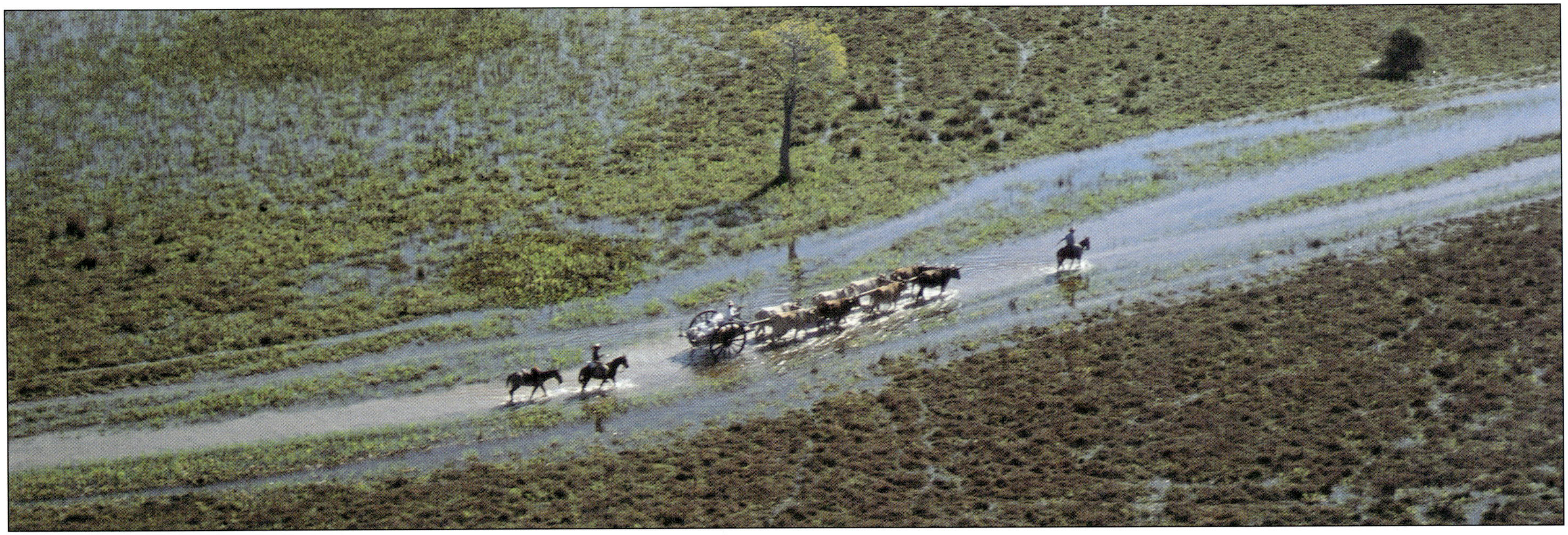

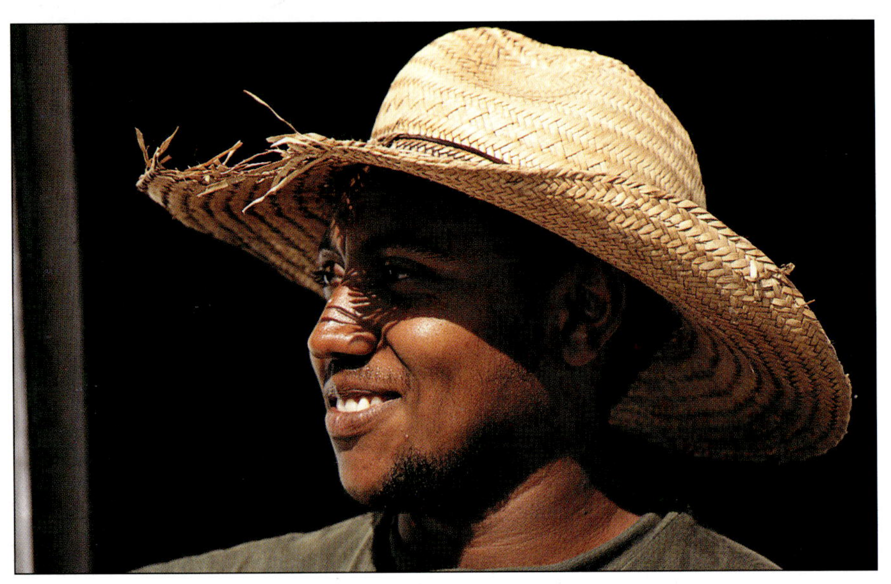

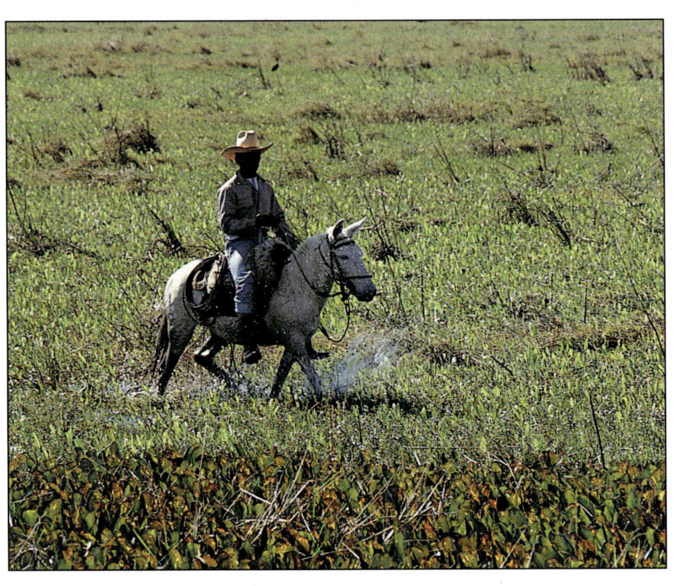

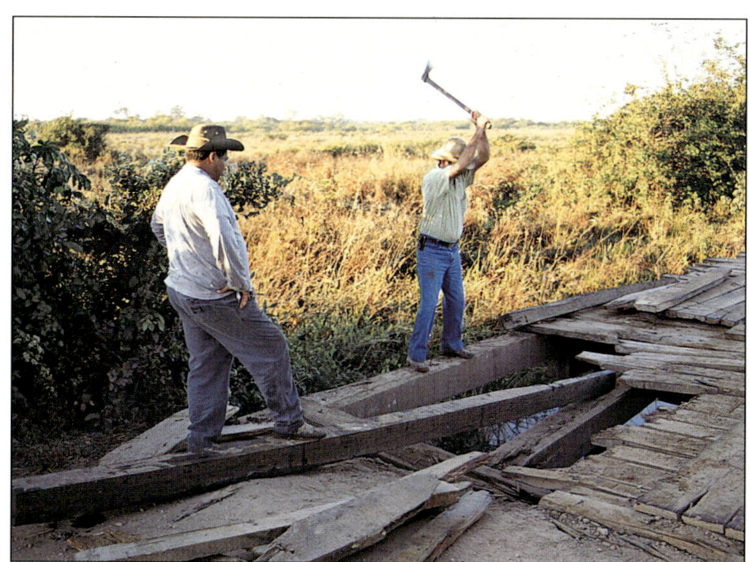

Repairing a Bridge

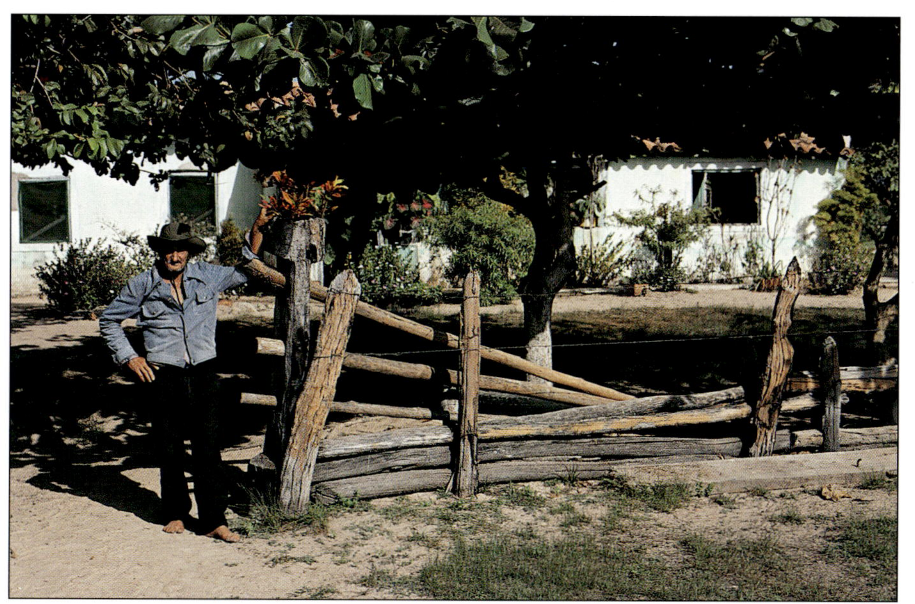

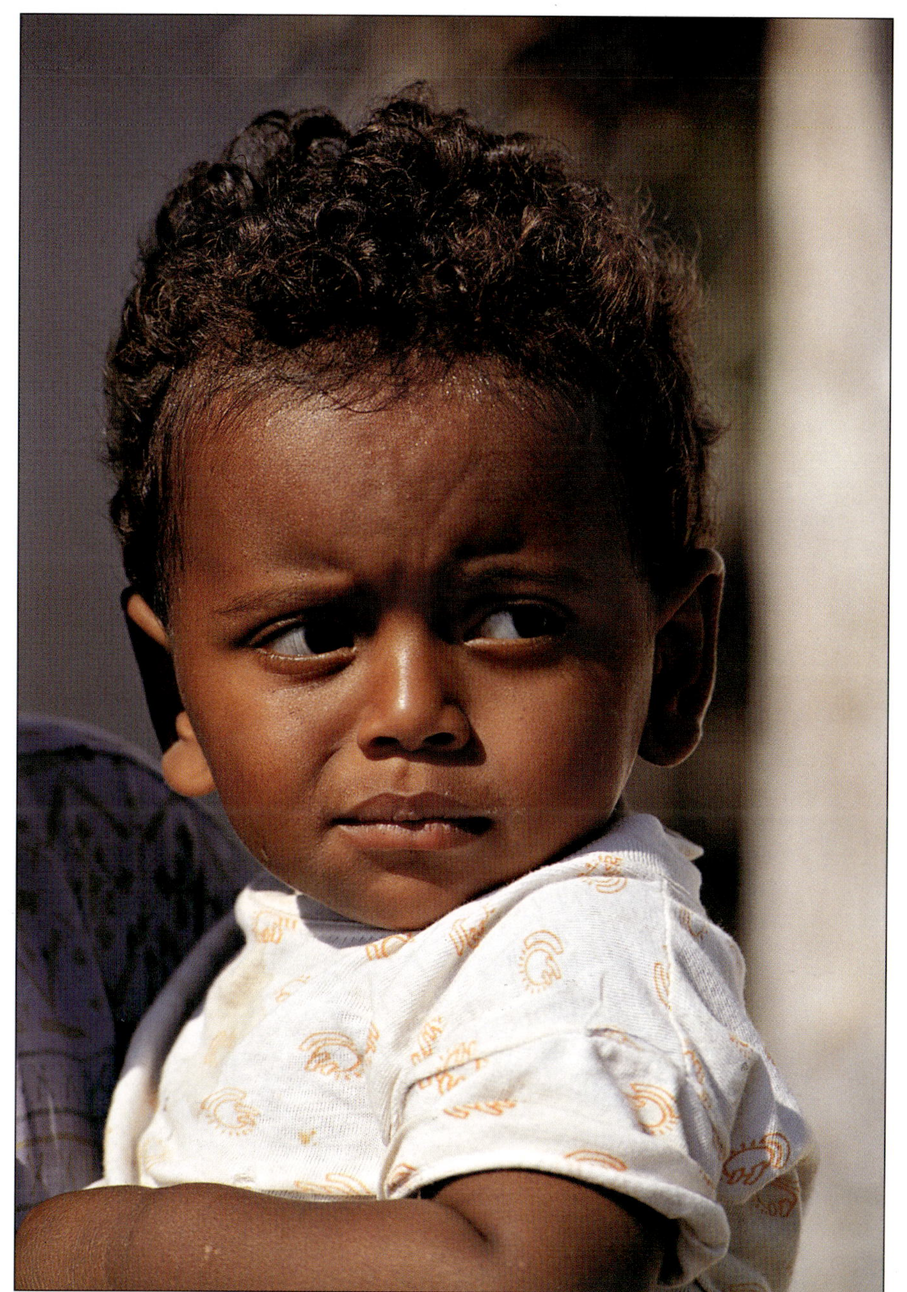

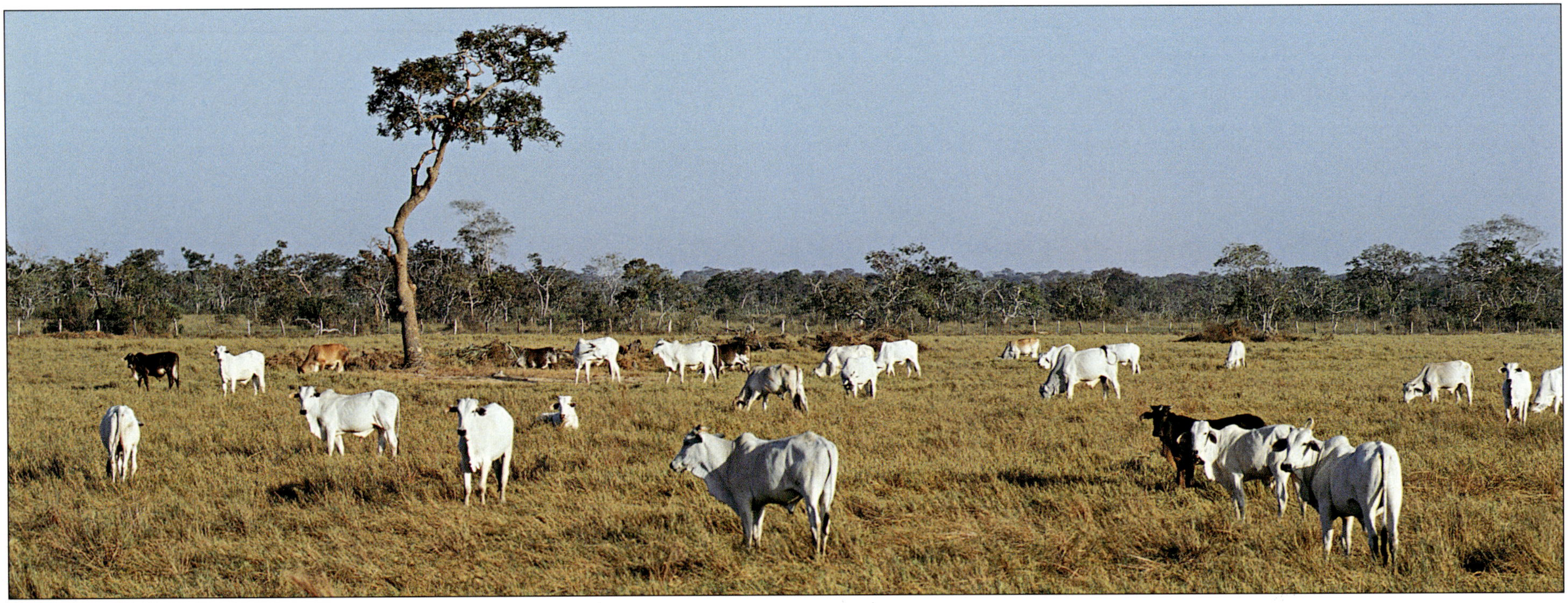

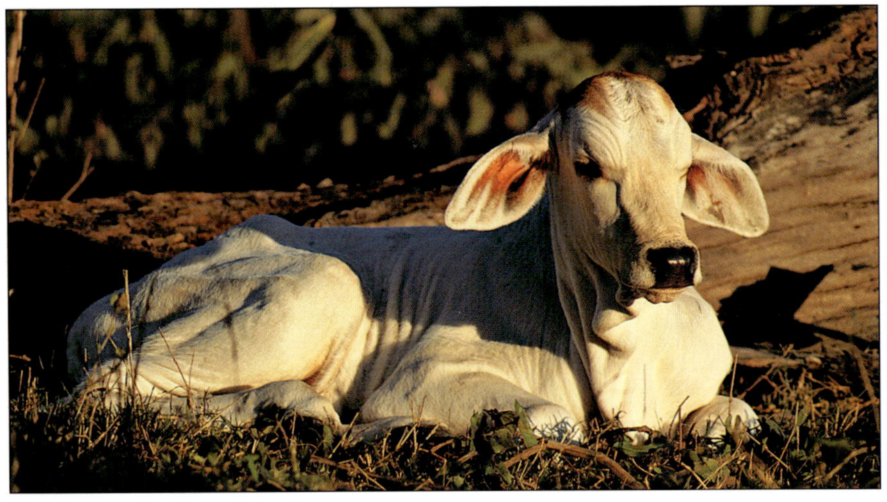

The history of cattle ranching goes back 250 years when immigrants from Europe invaded the area. They started in the north near Cuiaba and spread out from there. The ranches at that time were huge, encompassing up to 1,000,000 acres. After generations of dividing the land among siblings, most ranches today rarely exceed 25,000 acres. It is estimated in the earlier days that 700,000 cattle roamed the Pantanal. Today, calculations place that number at under five million. This number is held fairly constant because extensive flooding during the wet season limits the number of cattle the area can handle. The White Brahma thrives best in this environment and can readily be seen throughout the Pantanal.

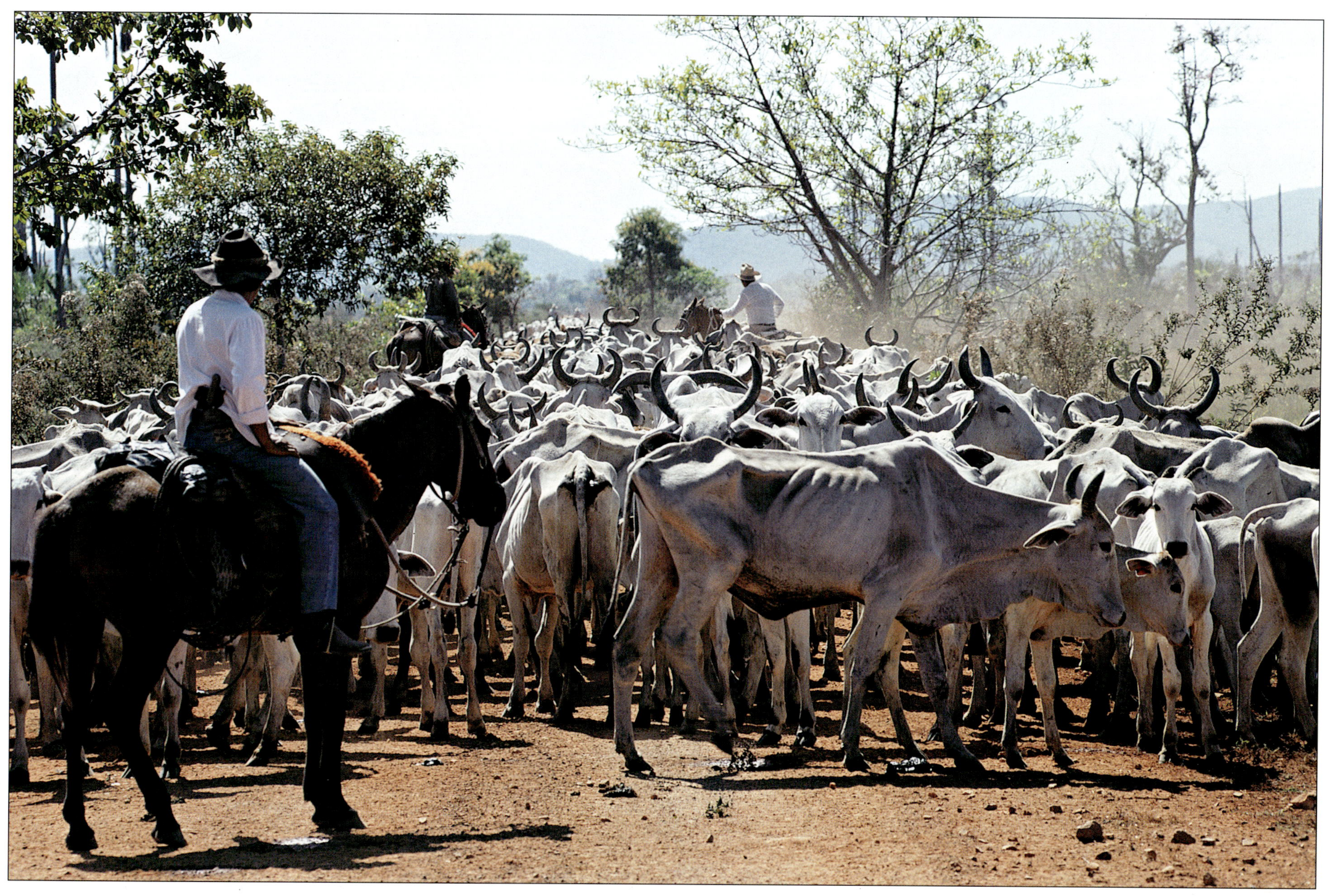

Rush Hour Traffic

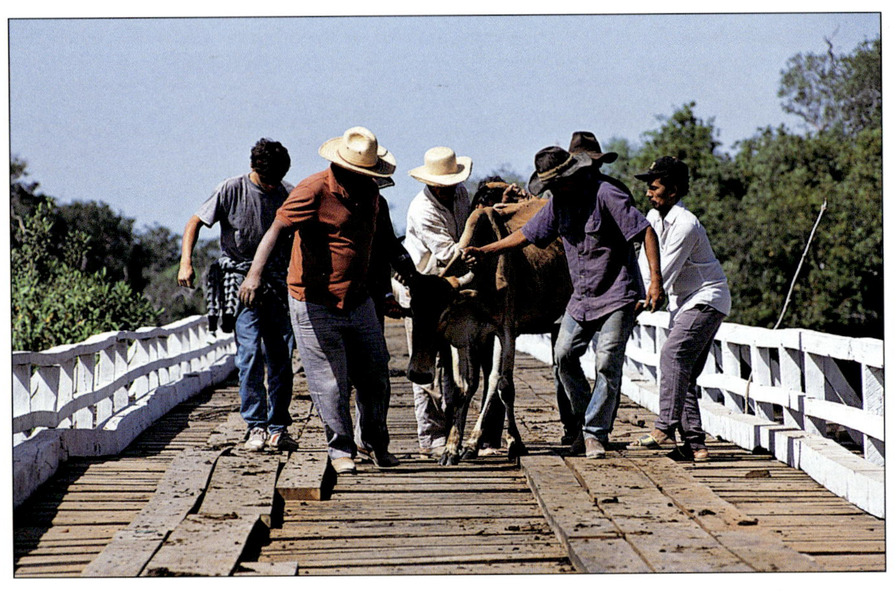

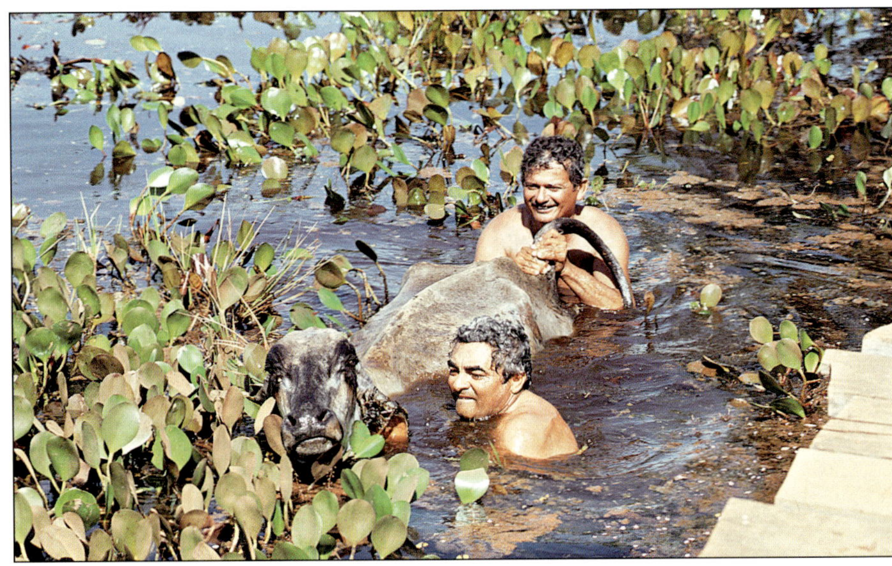

To control over-grazing, Pantaneiros, or wranglers, will drive cattle from one area of the ranch to another. The cattle have to cross bridges, which are not in the best condition. This is a scary situation for them as they try to keep track of four legs over a maze of holes. It is not uncommon for some cows being moved across a bridge to step into a hole or just sit down out of fear. These poor animals have to be picked up and carried off. Others will jump into the water in hopes of avoiding the bridge. In this situation the wrangler strips down and enters the water to manually push the cow to the other side. It is not as glamorous as it seems being a Pantaneiro in the Pantanal.

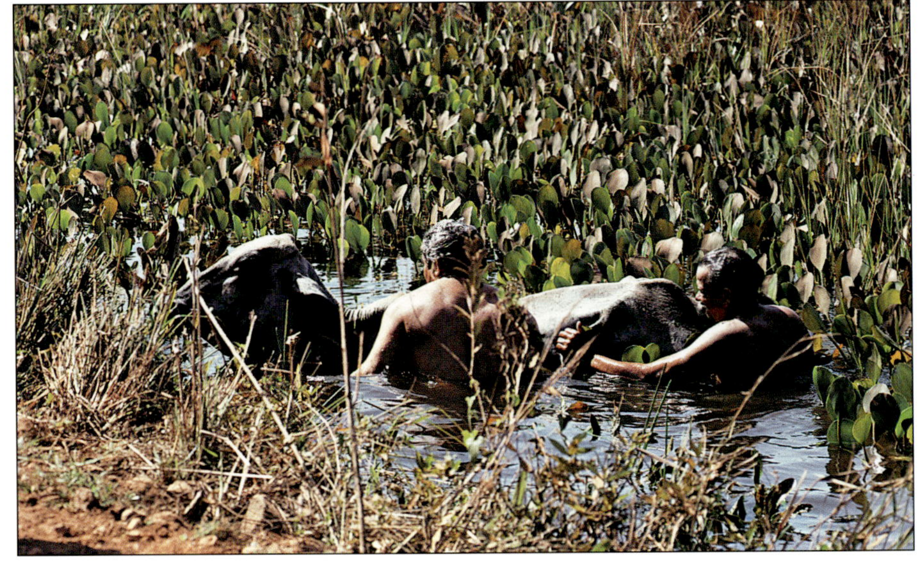

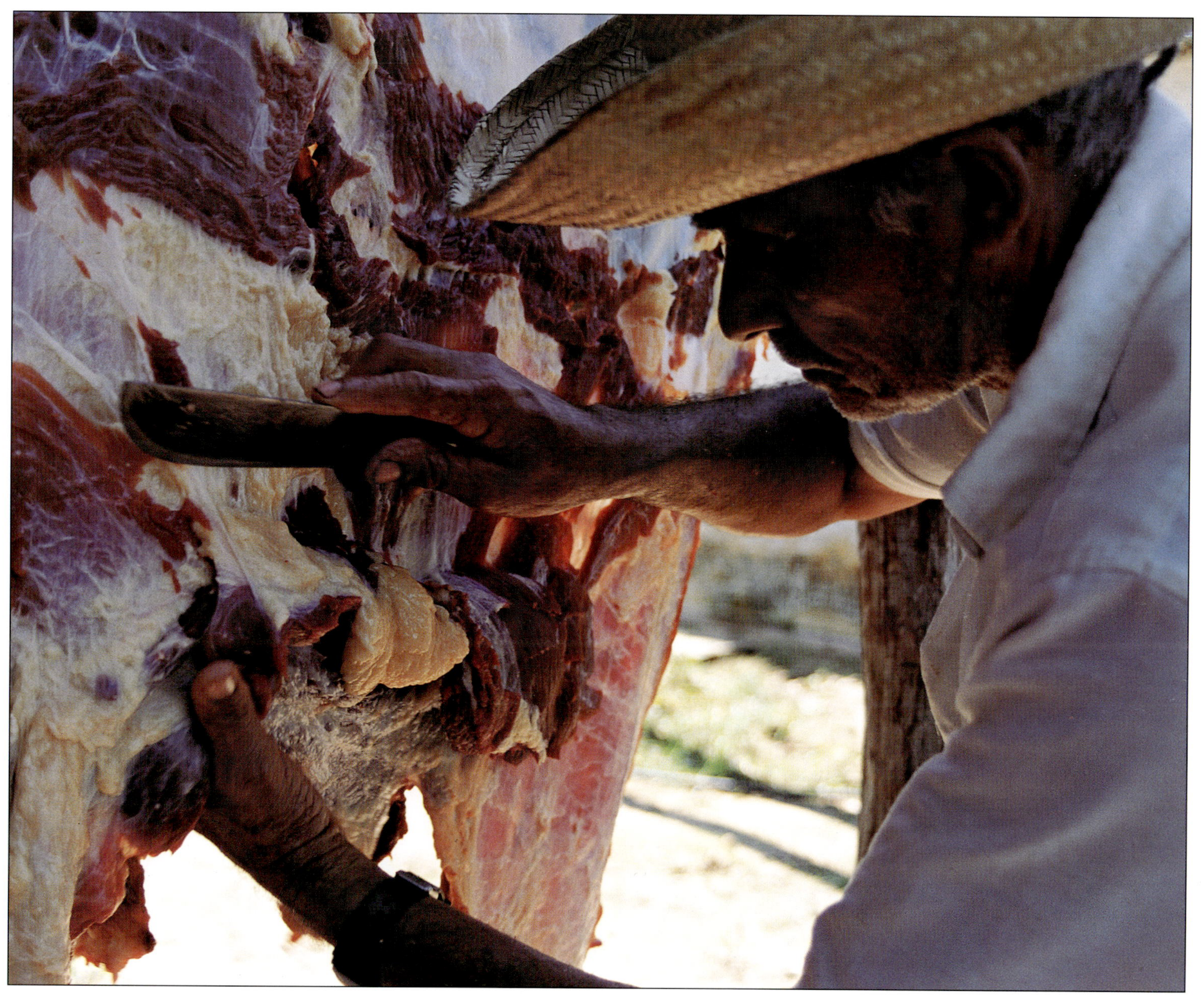

Cattle are often butchered for sustenance

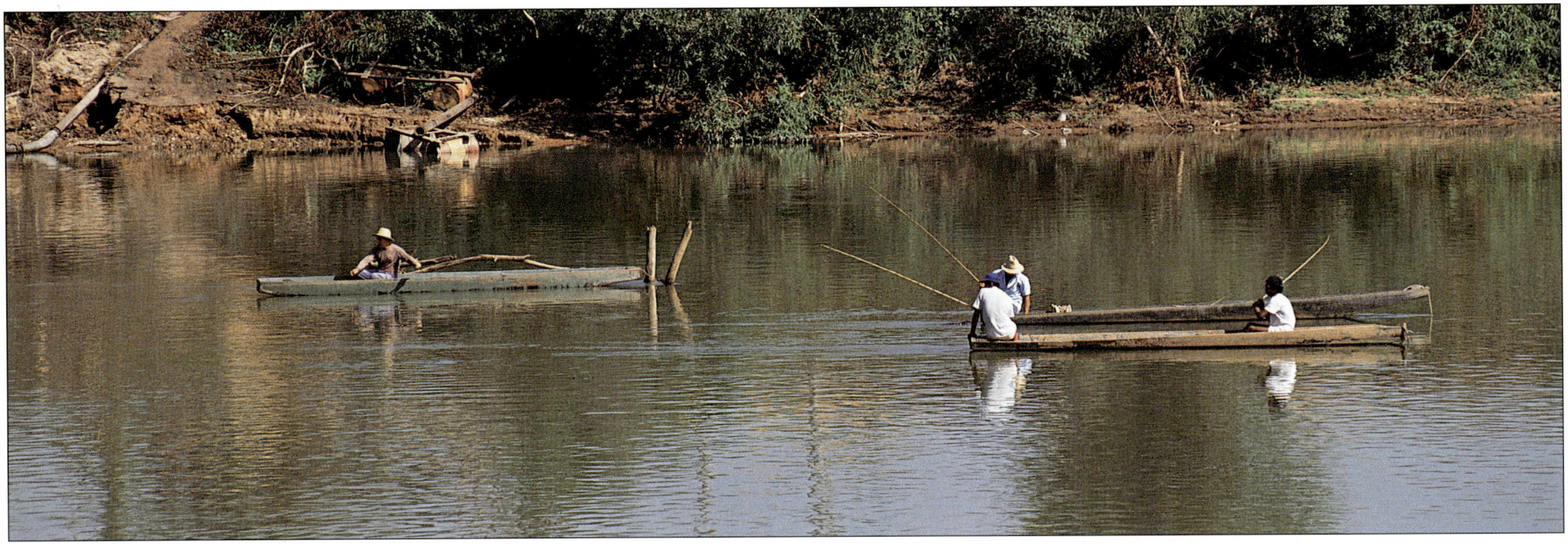

Locals fishing the Cuiaba River

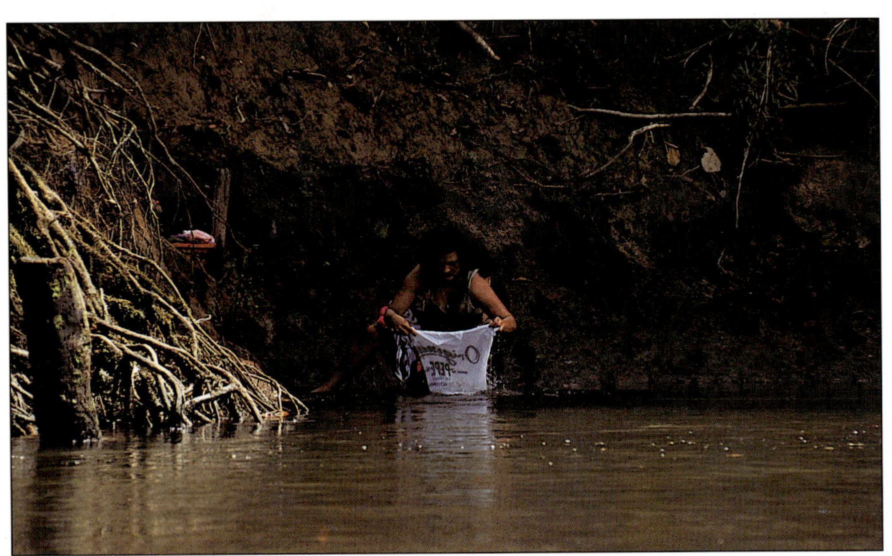

The Paraguay River and its tributaries are the lifeblood of the Pantanal. They are also home to some 263 species of fish. These fish are an important food source for both wildlife and local people. Due to remoteness of the Pantanal, people that live there fish because fish are an important part of their diet. Commerical fishermen also sell large volumes of these at local markets. Because of the immense volume of fish present, sport fishing is on the rise and pulls in ever increasing numbers of fishermen from around world. How much longer this resource can take such pressure is a worthy question.

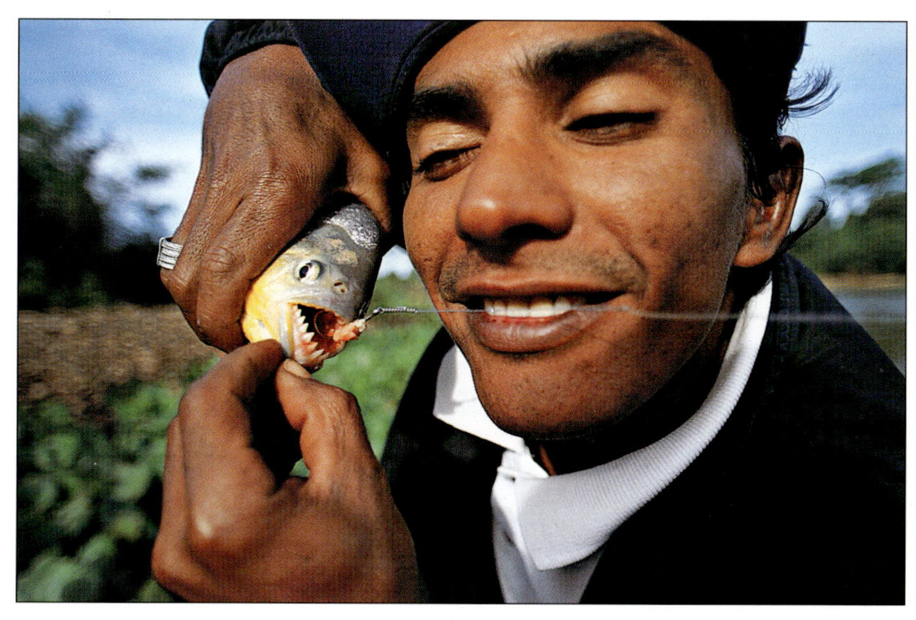

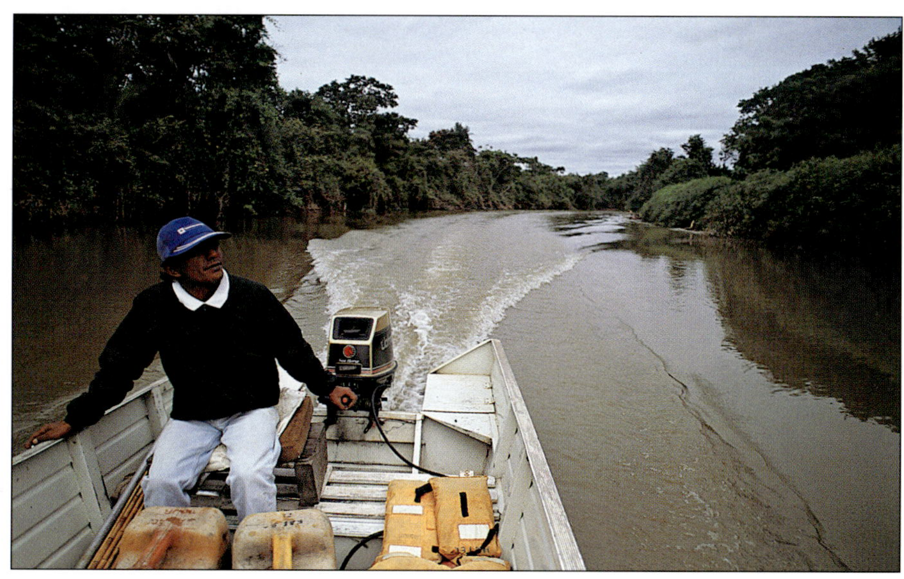

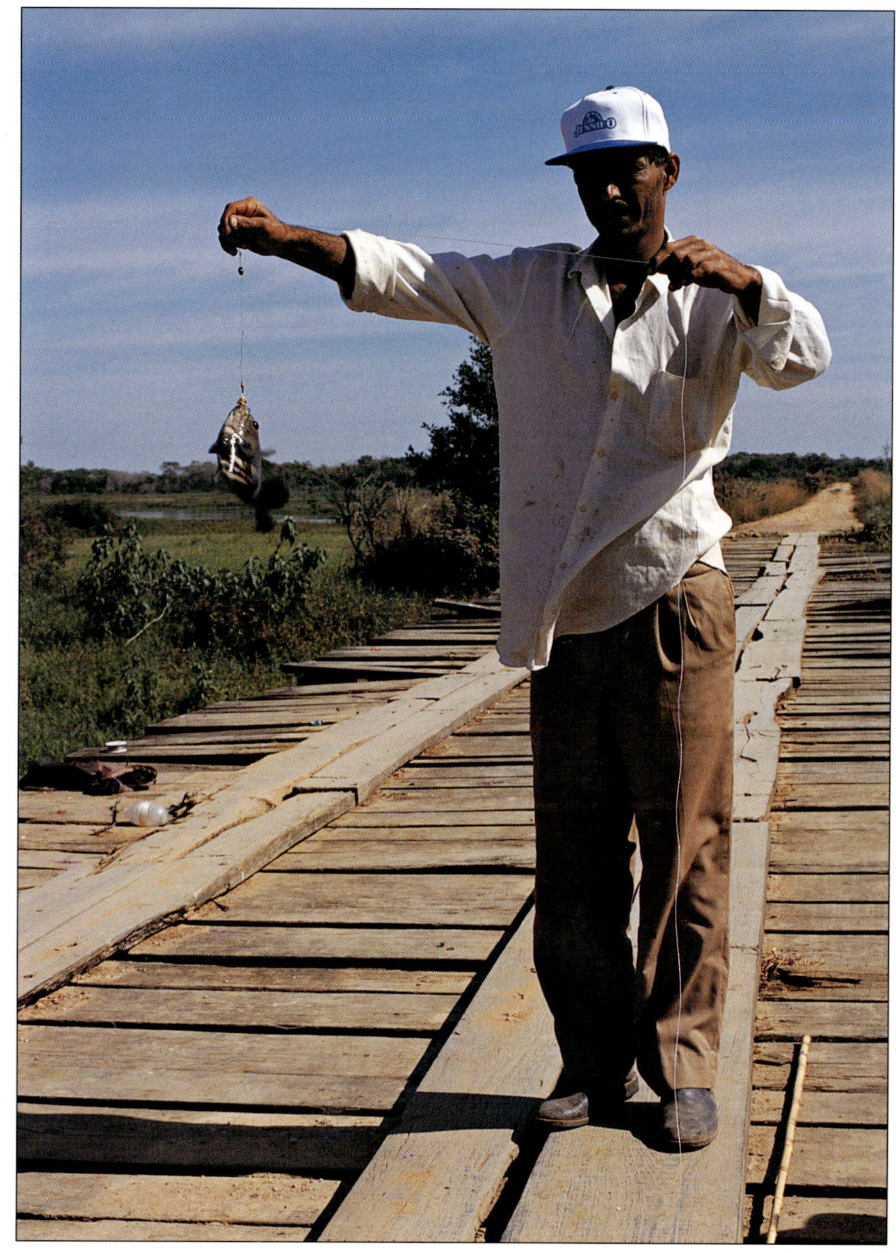

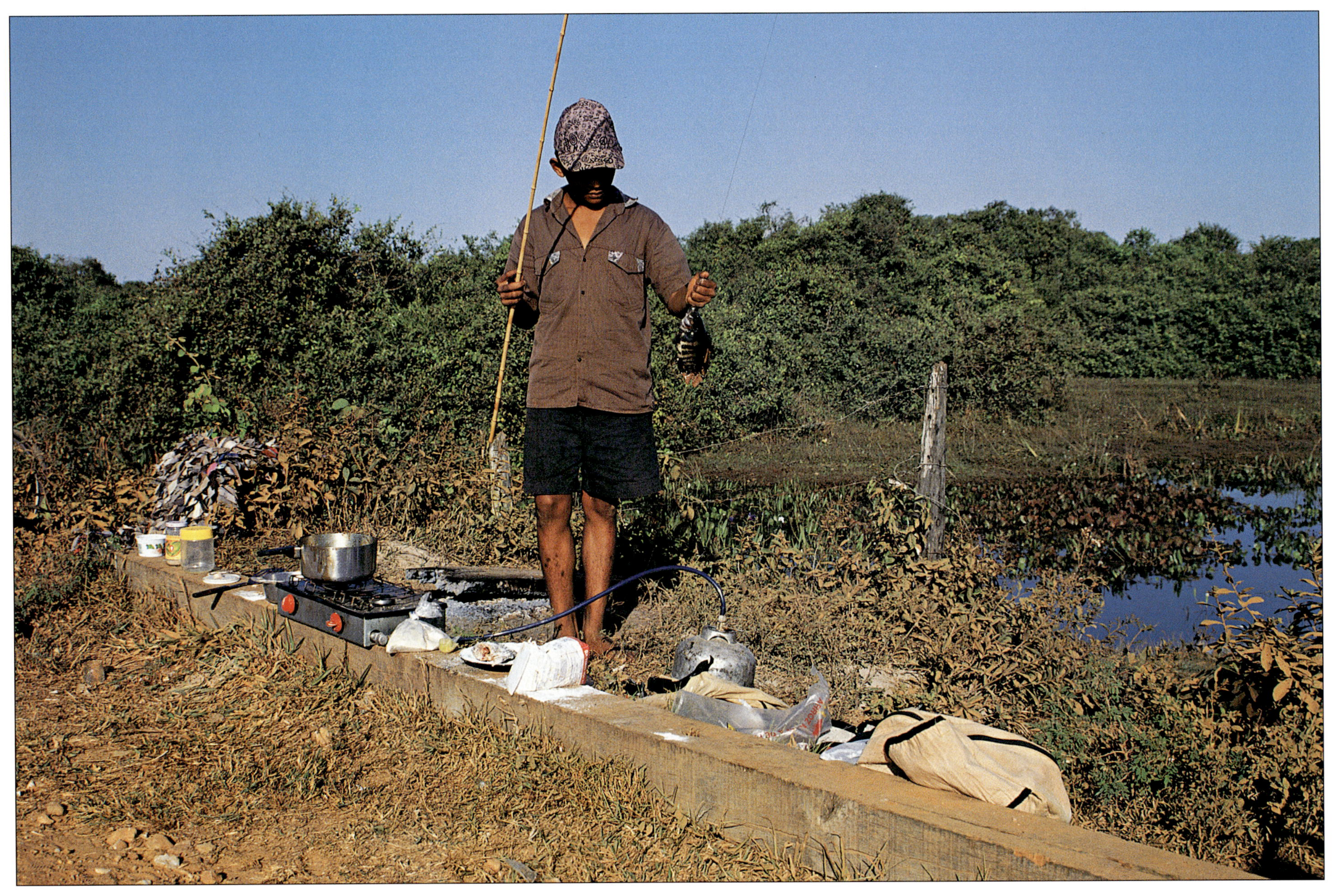

Young fisherman along the Transpantaneria

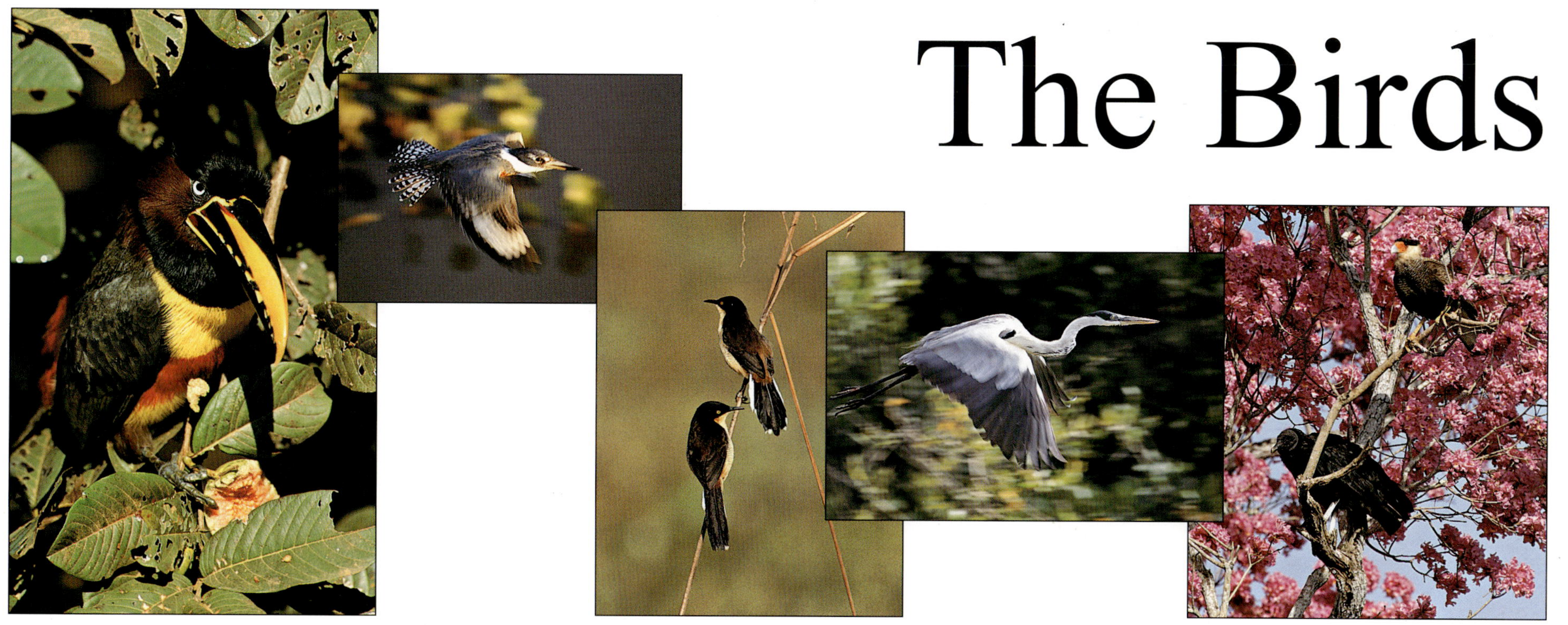

The Birds

There is a saying, "If there is something to eat—there is something there to eat it." How true this becomes in relation to the Pantanal, especially with its avian population. The birds are there because the biomass, or food production, can sustain them. Close to 700 species are found there. Just to get a feeling of this amount, the Pantanal is about the size of Texas. All of North America has a little over 700 species.

Not all of the birds, such as the Snail Kite or Great Egret, are indigenous or year-round resident. Some come for a visit from not-so-far-away locations such as Argentina and Chile. Others migrate from a much longer distance like North America. It is thought the Pantanal has the distinction of having three major South American migratory routes.

The large food production can also be responsible in that the area boasts some of the larger members of many groups in the world. The world's largest stork, macaw and toucan all reside there. The largest kingfisher in the Americas, the Ringed Kingfisher, can be found in staggering numbers.

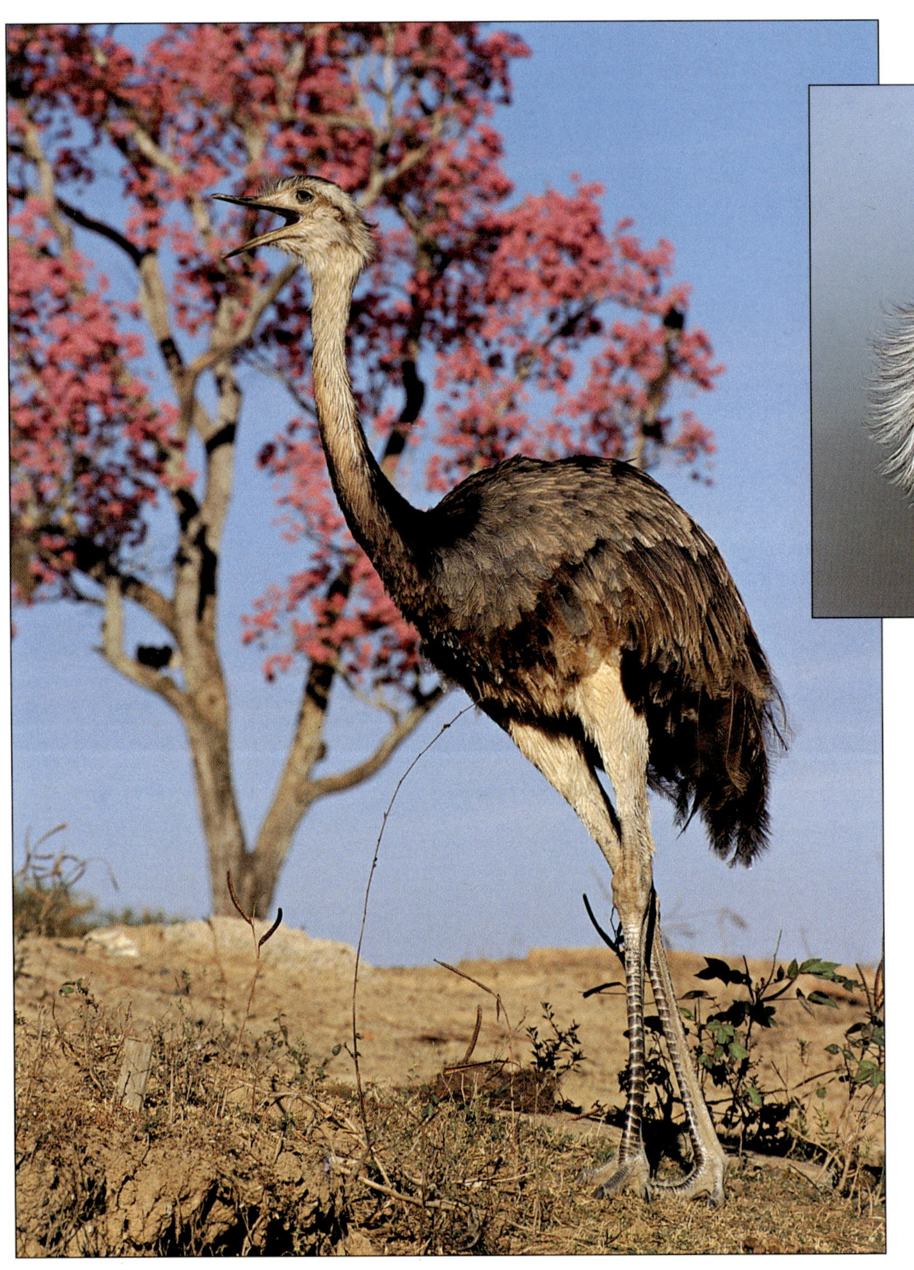

Greater Rhea
Rhea americana

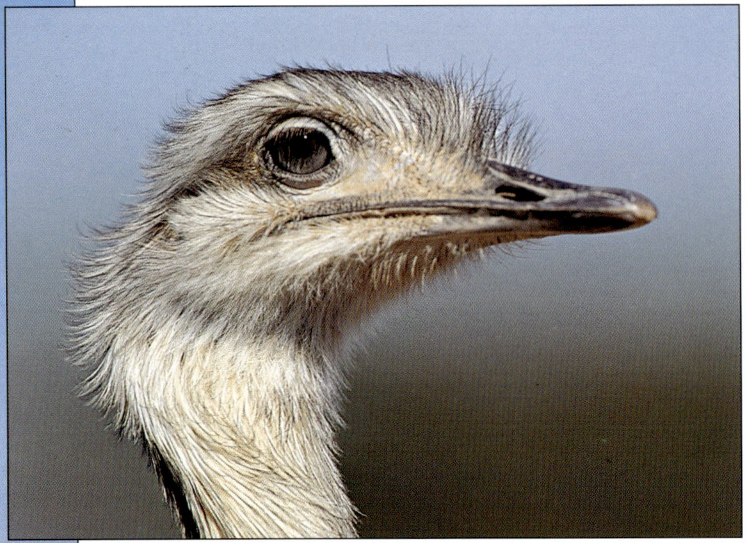

The Rhea is a member of a group called ratites or flightless birds. In the long evolution of ratites, they failed to develop the median ridge, or keel, on the breastbone. Without this extra surface area for muscle attachment these birds are rendered flightless. When threatened the rhea can do one of two things. It will run to escape. Or, it will try to avoid detection by sitting with its neck outstretched along the ground.

Unlike most avian species, it is the male's job to incubate the nest with eggs. Hatched eggs are also raised under his care. The way this comes about is that a territory is established by the male who attracts and mates with several females. Each hen deposits eggs in the nest and it is the cock's job to incubate these eggs and raise the young. The hen immediately leaves and mates with other males to contribute her eggs to their nests.

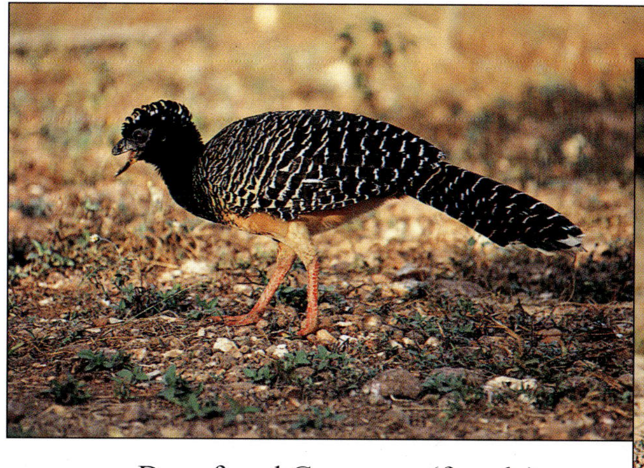

Bare-faced Curassow (female)

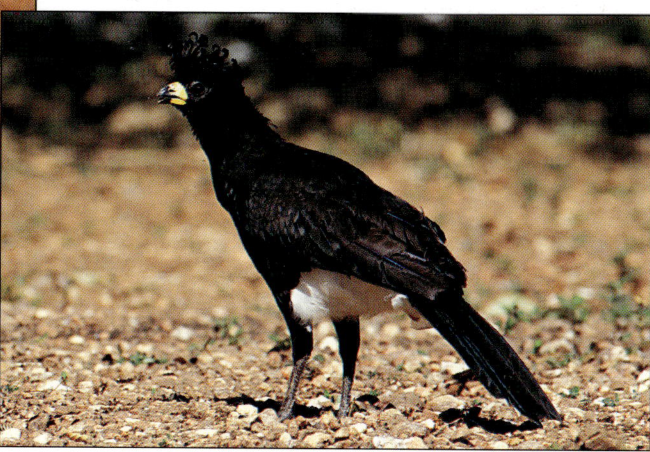

Bare-faced Curassow (male)
Crax fasciolata

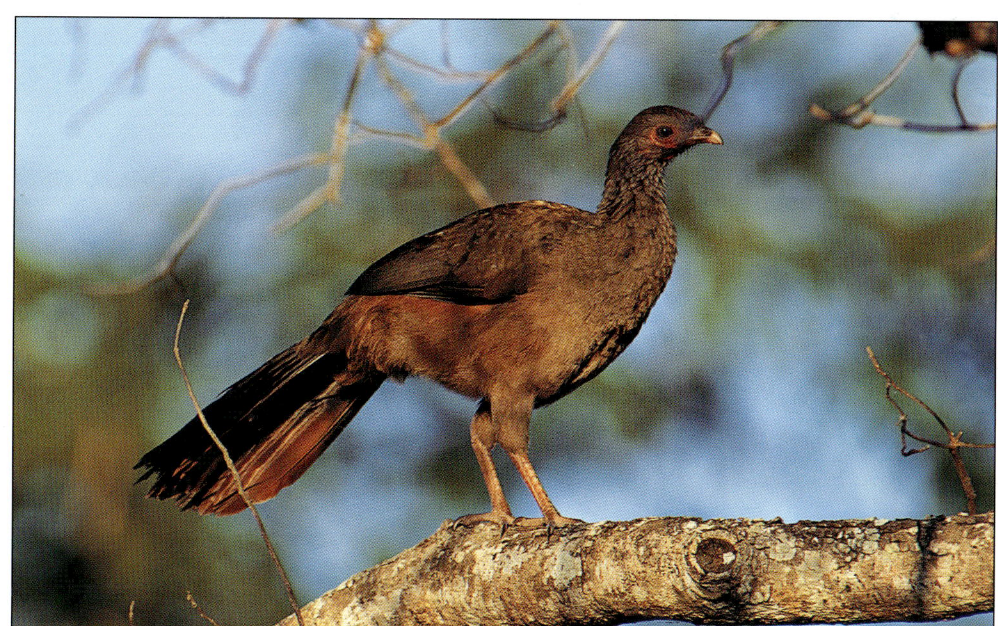

Chaco Chachalaca
Ortalis canicollis

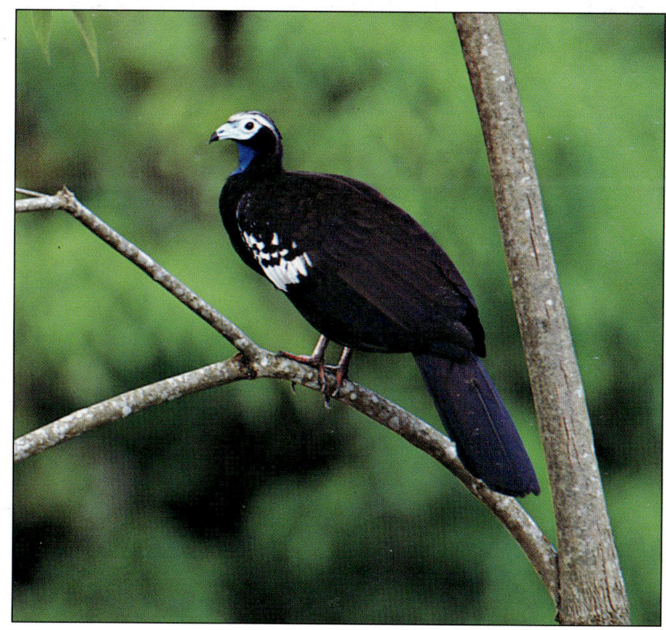

Blue-throated Piping Guan
Aburris pipile

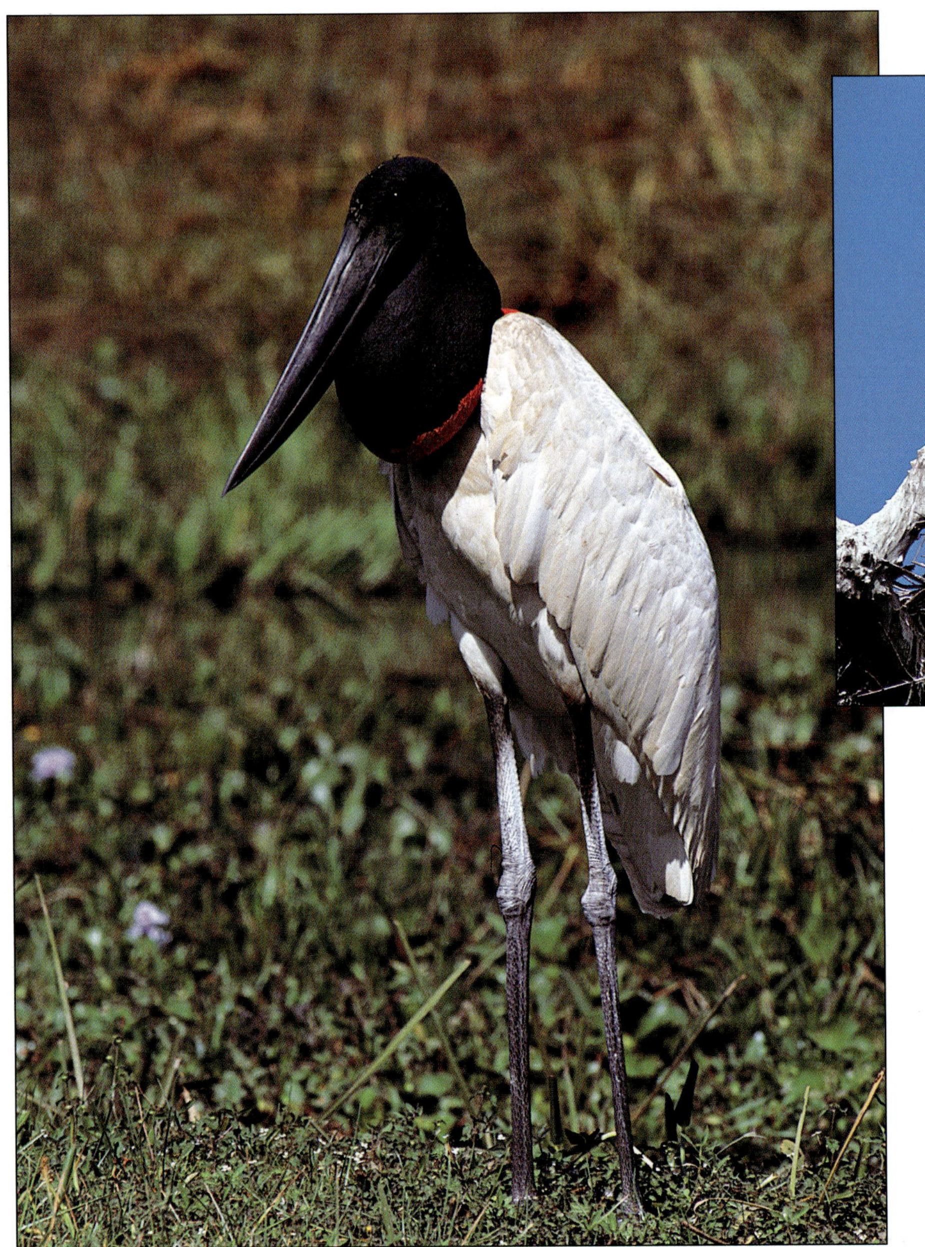

Jabiru
Jabiru mycteria

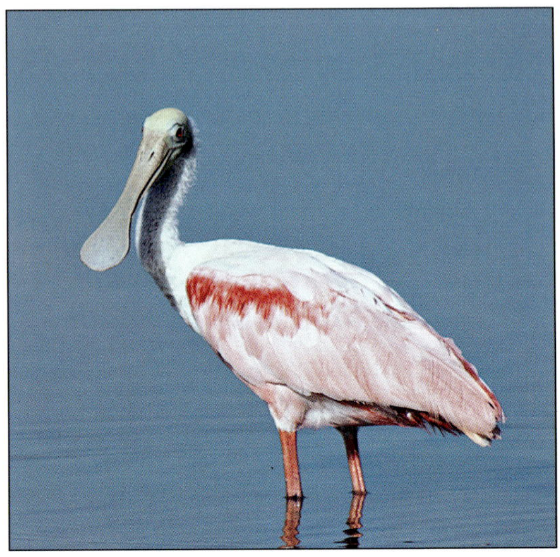

Roseate Spoonbill
Ajaia ajaja

There is one bird often considered the mascot of the Pantanal. It is the Jabiru stork. At close to five feet tall, it is the biggest stork in the world. Readily seen foraging for critters along the Transpantaneria, it is strict carnivore. Fish, snails, frogs, large insects or even snakes and small caiman fall prey to the Jabiru. The stork's large stick nest, a meter across, is used year after year by the same pair. Each year new sticks are added to the nest. This remodeling bulks up the nest. Monk parakeets often make their home in the lower portions, forming a symbiotic relationship with the Jabiru. For their free rent, the parakeets warn the Jabiru by becoming chattery when intruders come near the nest.

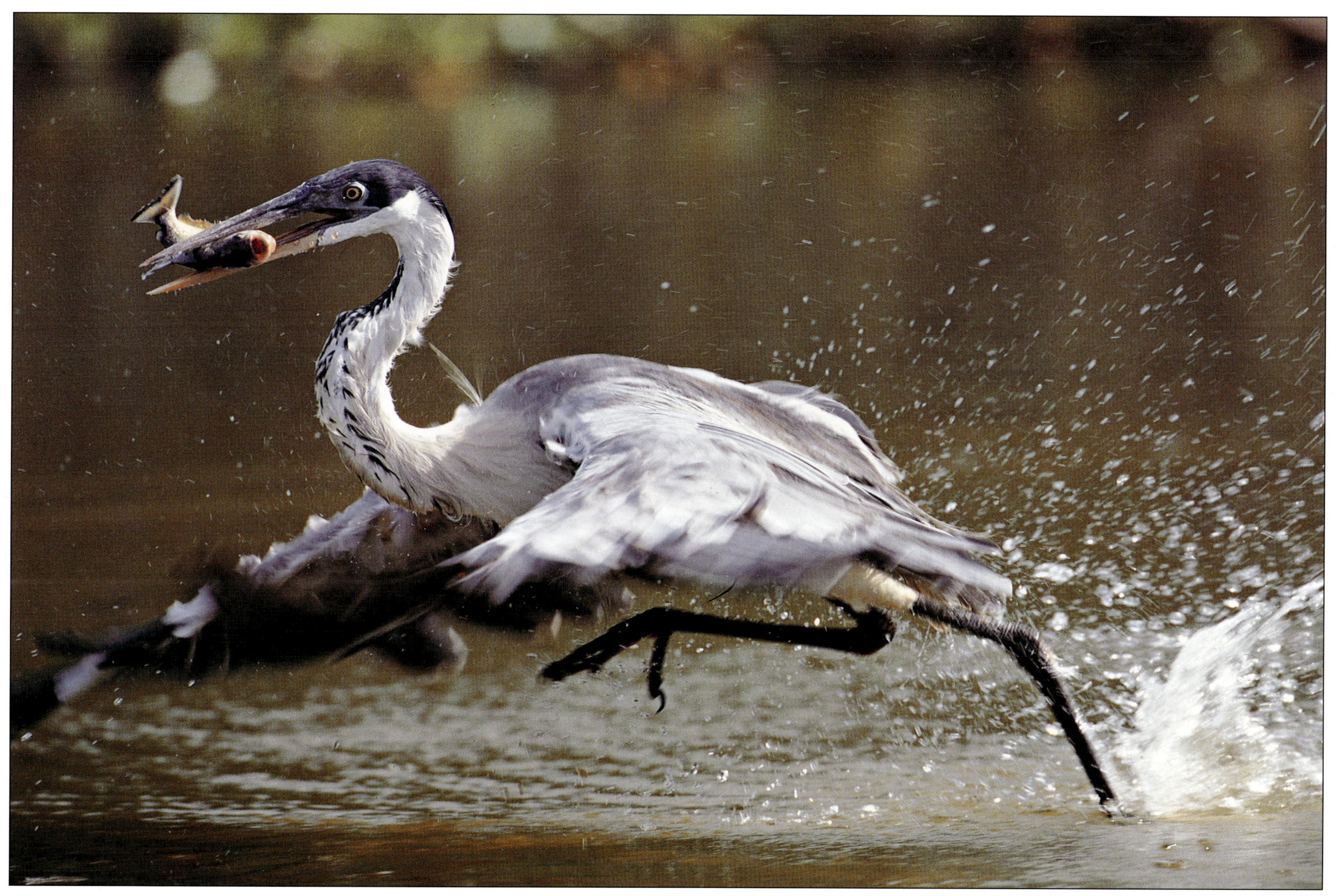

White-necked Heron
Ardea cocoi

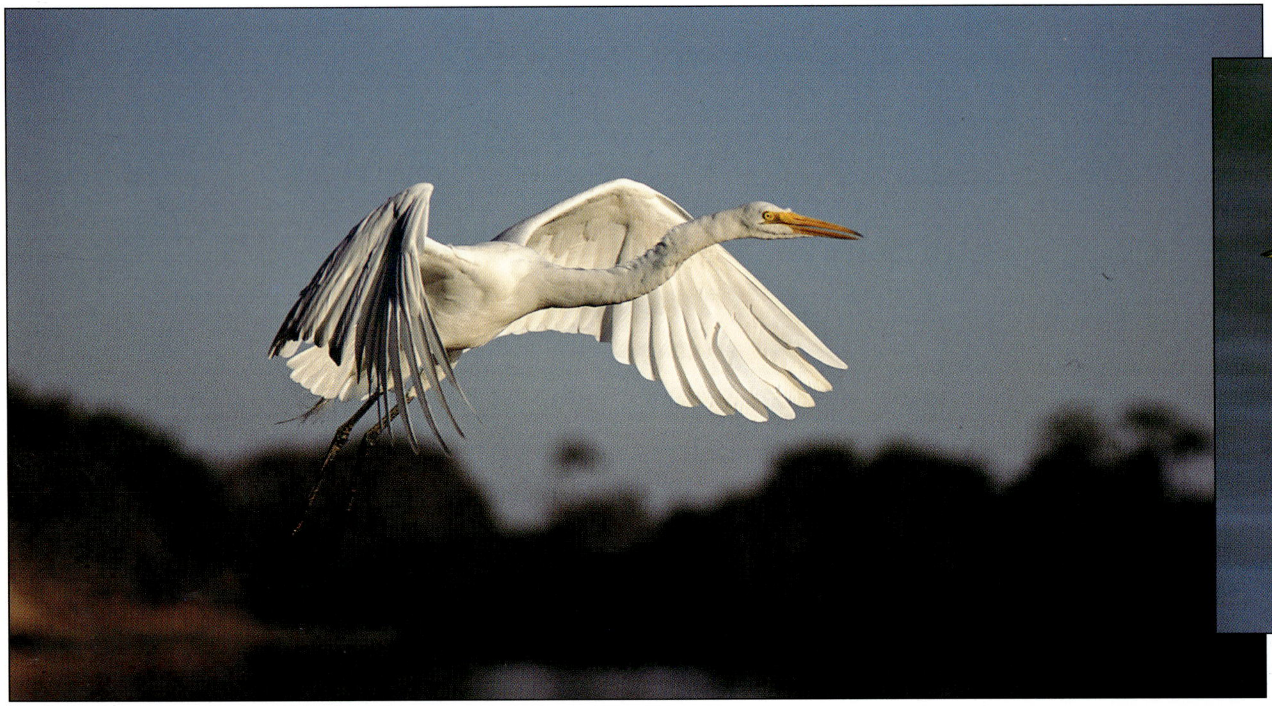

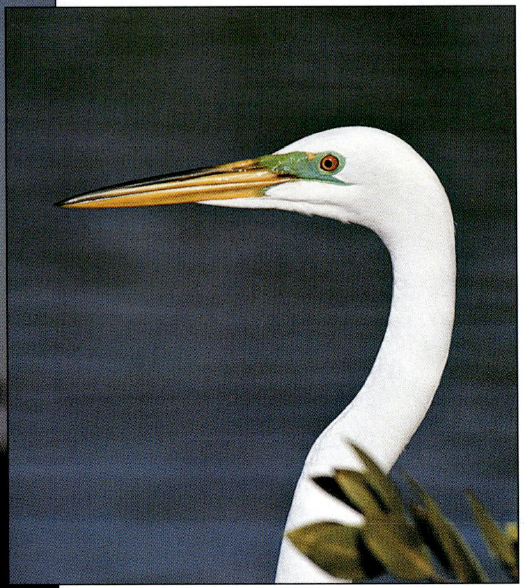

Great Egret
Casmerodius albus

Maguari Stork
Euxenura maguari

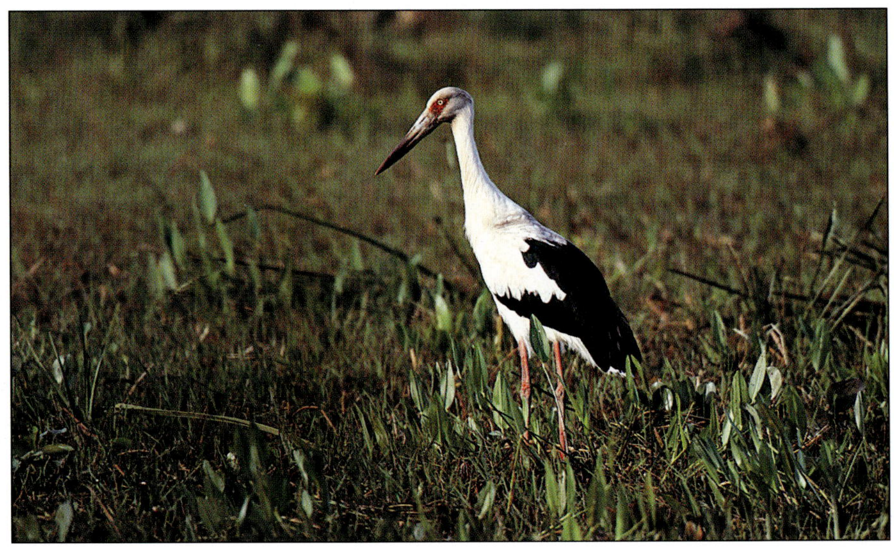

The Great Egret has to be the most numerous avian species found in the Pantanal. Be careful not to confuse it with the Snowy Egret or Cattle Egret which are also found in generous numbers. An easy way to remember differences is that the Great Egret has a yellow bill and black legs. The Snowy Egret has a black bill and black legs with bright yellow feet. Our third white species, the Cattle Egret, has a yellow bill and yellow legs.

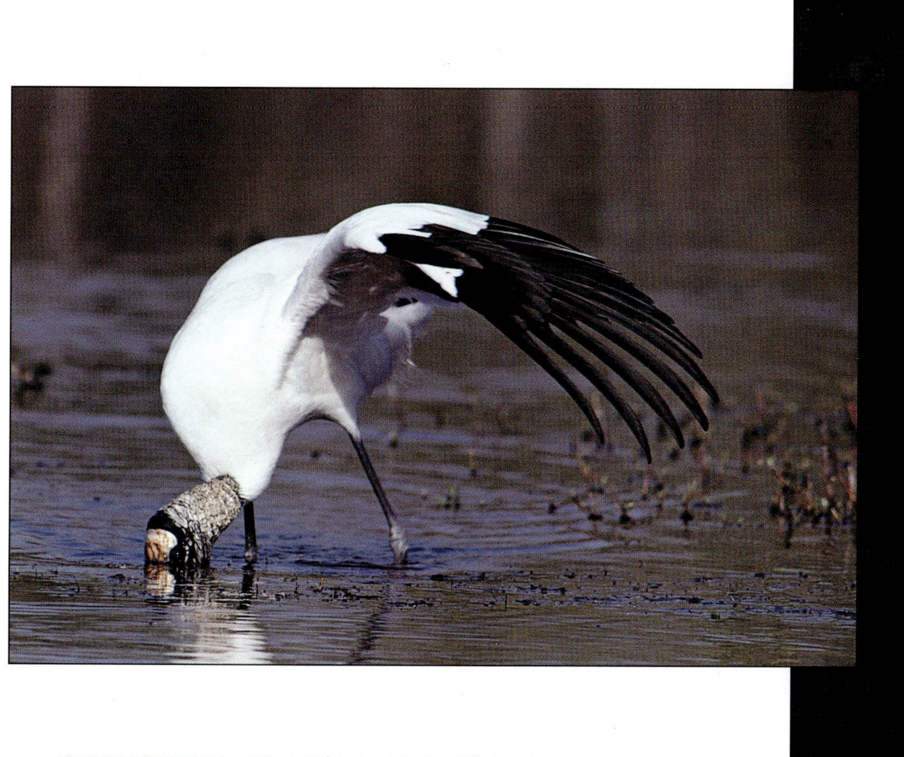

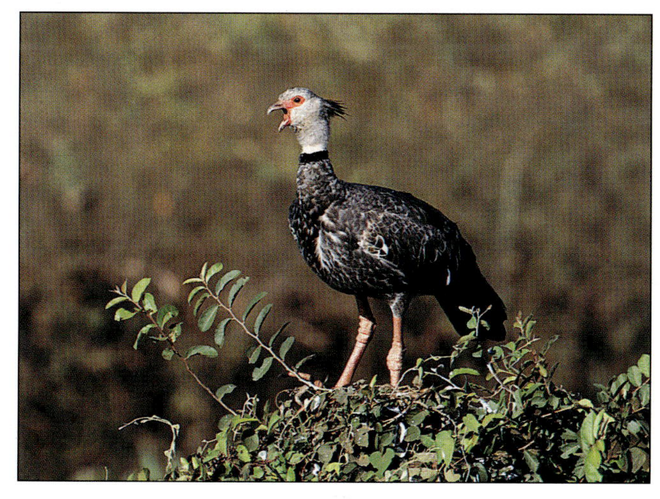

Southern Screamer
Chauna torquata

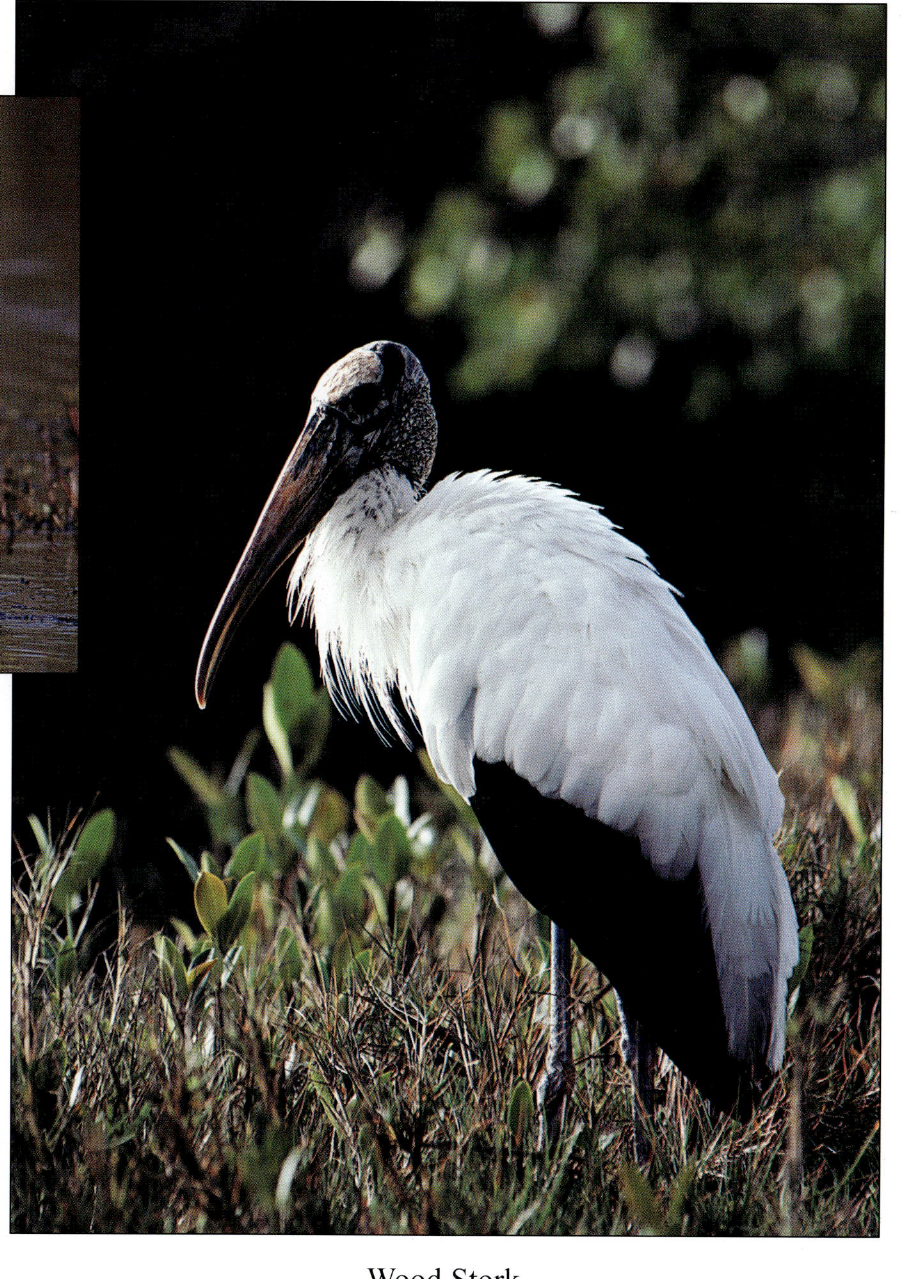

Wood Stork
Mycteria americana

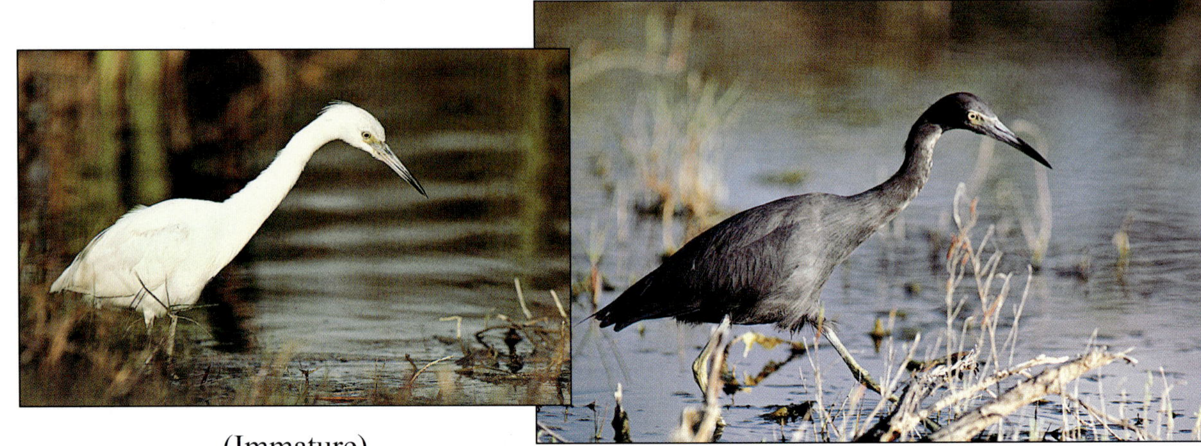

(Immature)

Little Blue Heron
Egretta caerulea

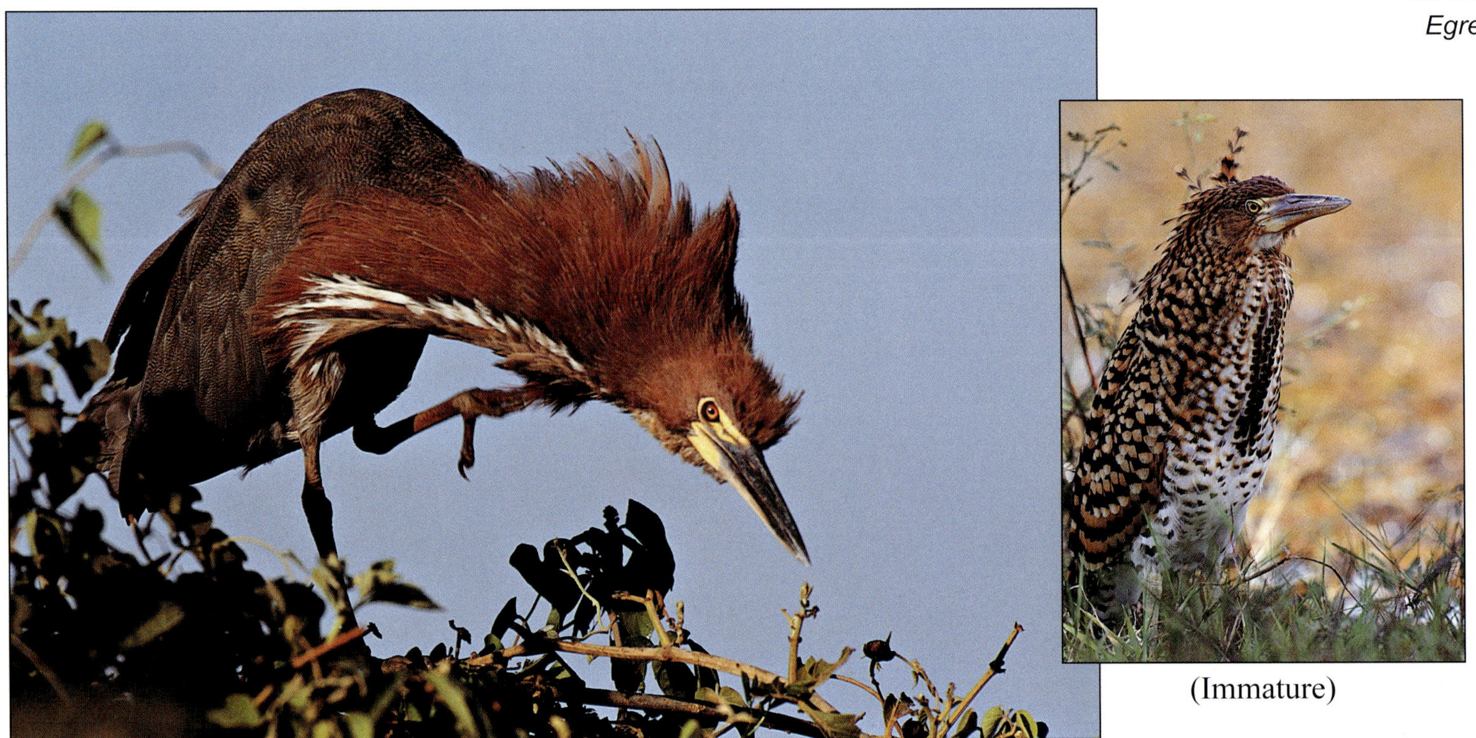

(Immature)

Rufescent Tiger-Heron
Tigrisoma lineatum

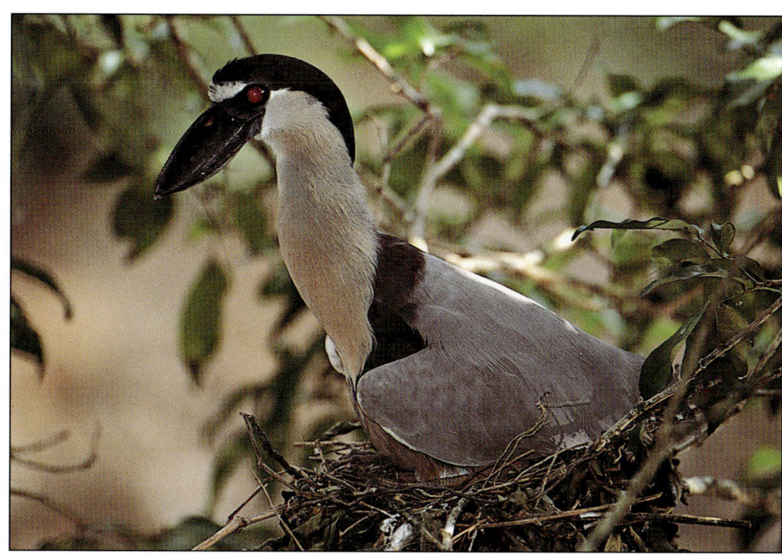

Boat-billed Heron
Cochlearius cochlearius

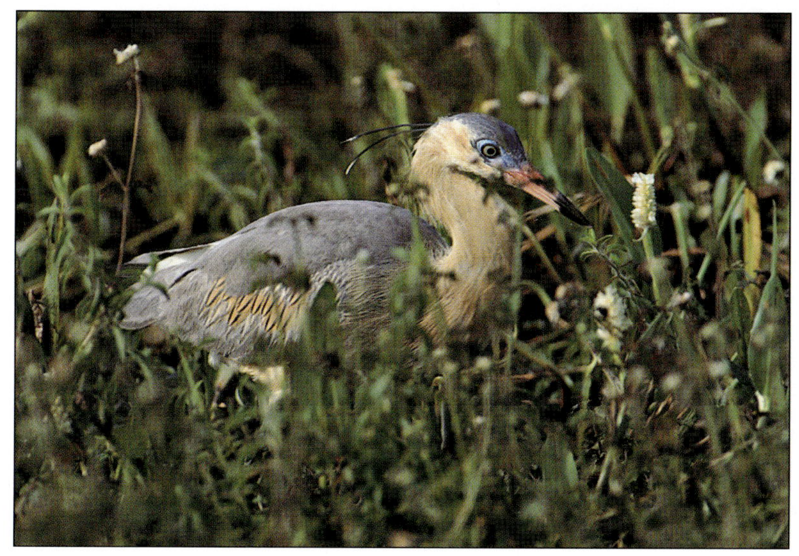

Whistling Heron
Syrigma sibilatrix

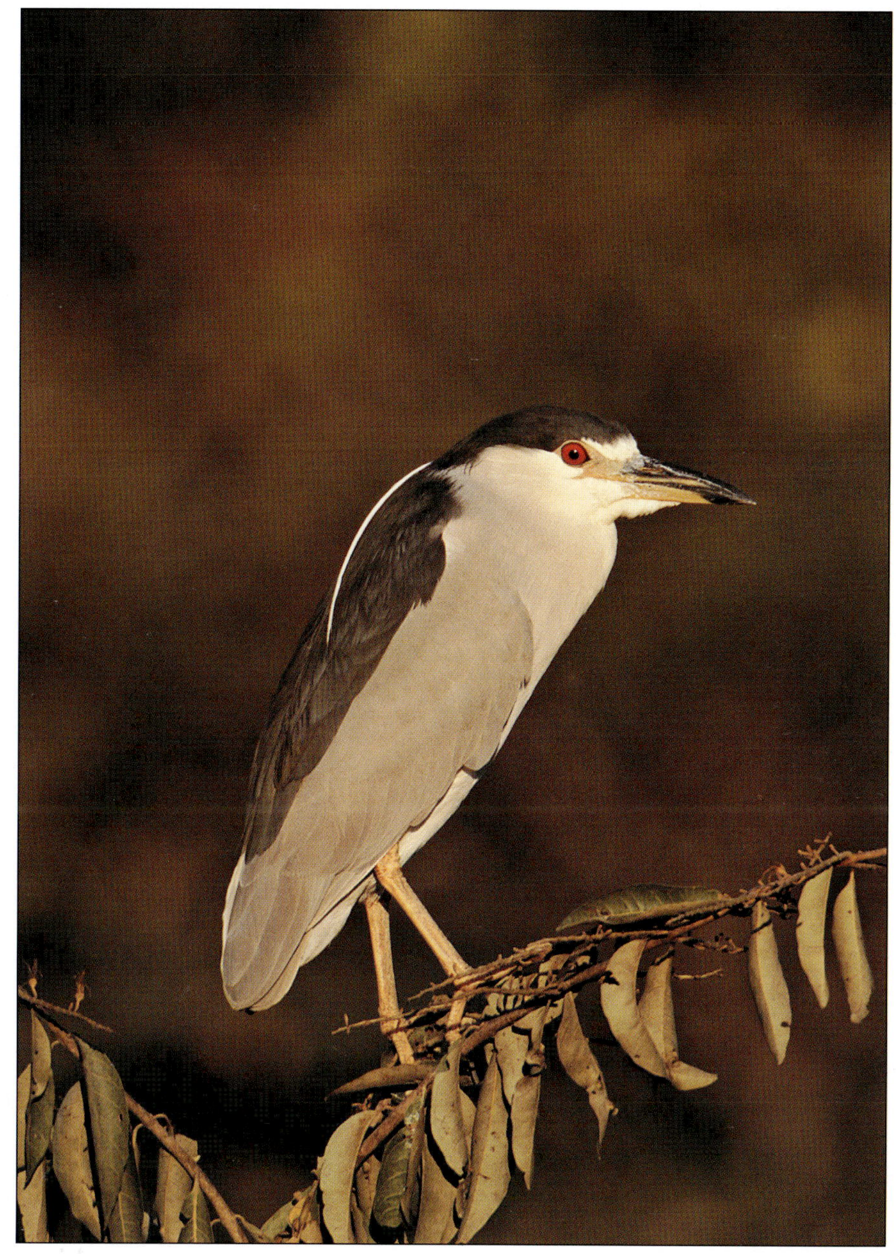

Black-crowned Night Heron
Nycticorax nycticorax

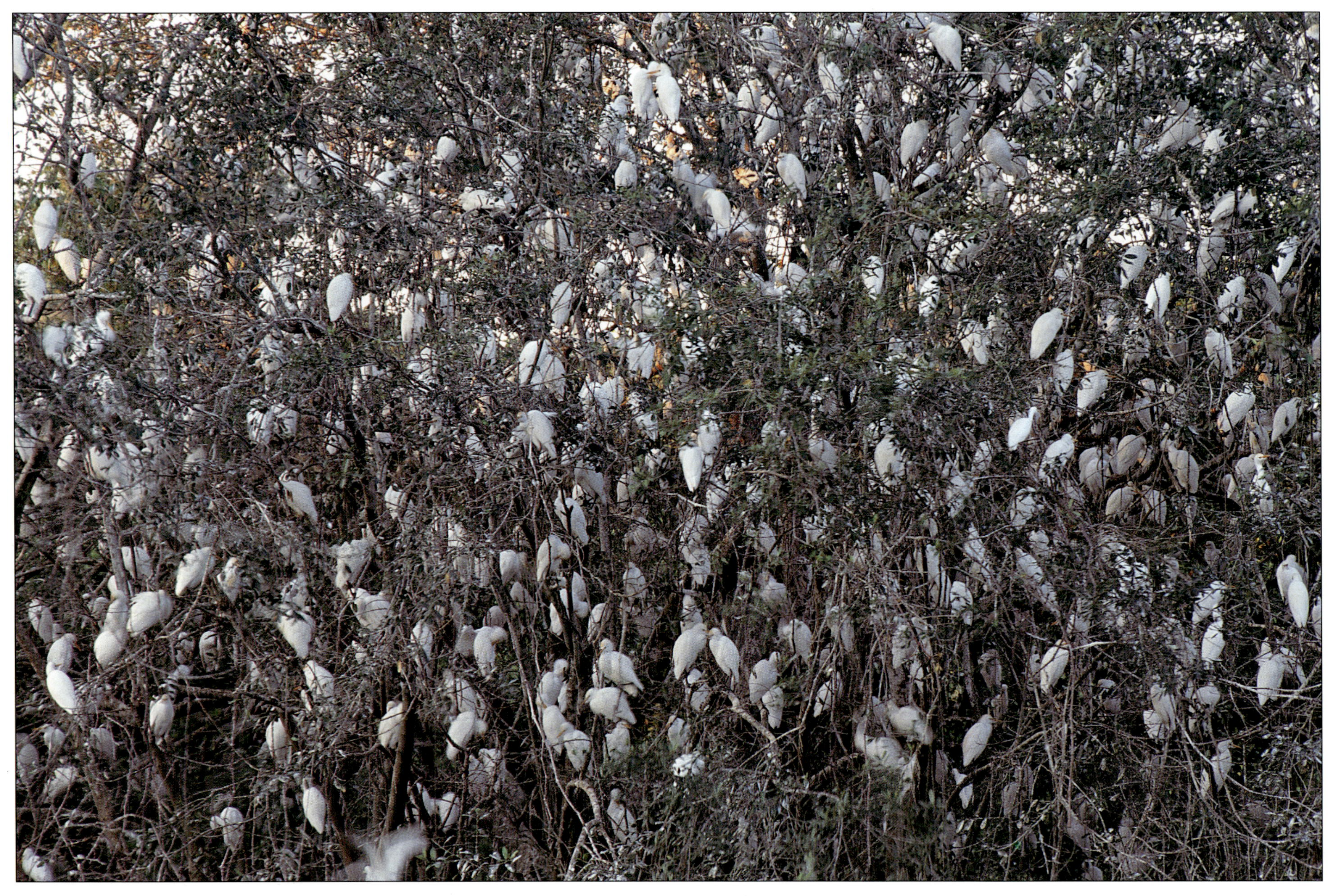

Cattle Egrets Roosting
Bubulcus ibis

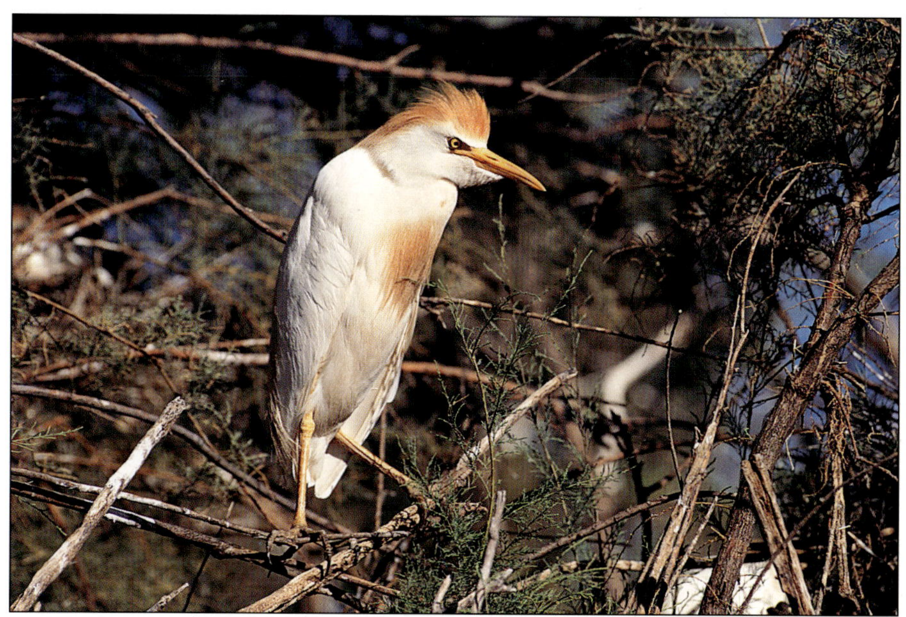

Cattle Egret (Breeding plumage)
Bubulcus ibis

It is thought Cattle Egrets self-introduced themselves to South America in the late 1800's by flying across the Atlantic from Africa. This bird was not seen in the Pantanal until the mid 1970's. With its characteristic behavior of sticking close to cattle to procure food, this egret doesn't compete with native species. The little niche they fill is to pick off insects, mostly grasshoppers, which are frightened from the grass as cattle walk or graze. Anywhere from one to seven or eight of these birds will accompany a cow, some of which occasionally perch atop their host. Other social behaviors exhibited by this bird are nesting in rookeries and using one tree as a communal roost for the night during non-breeding season.

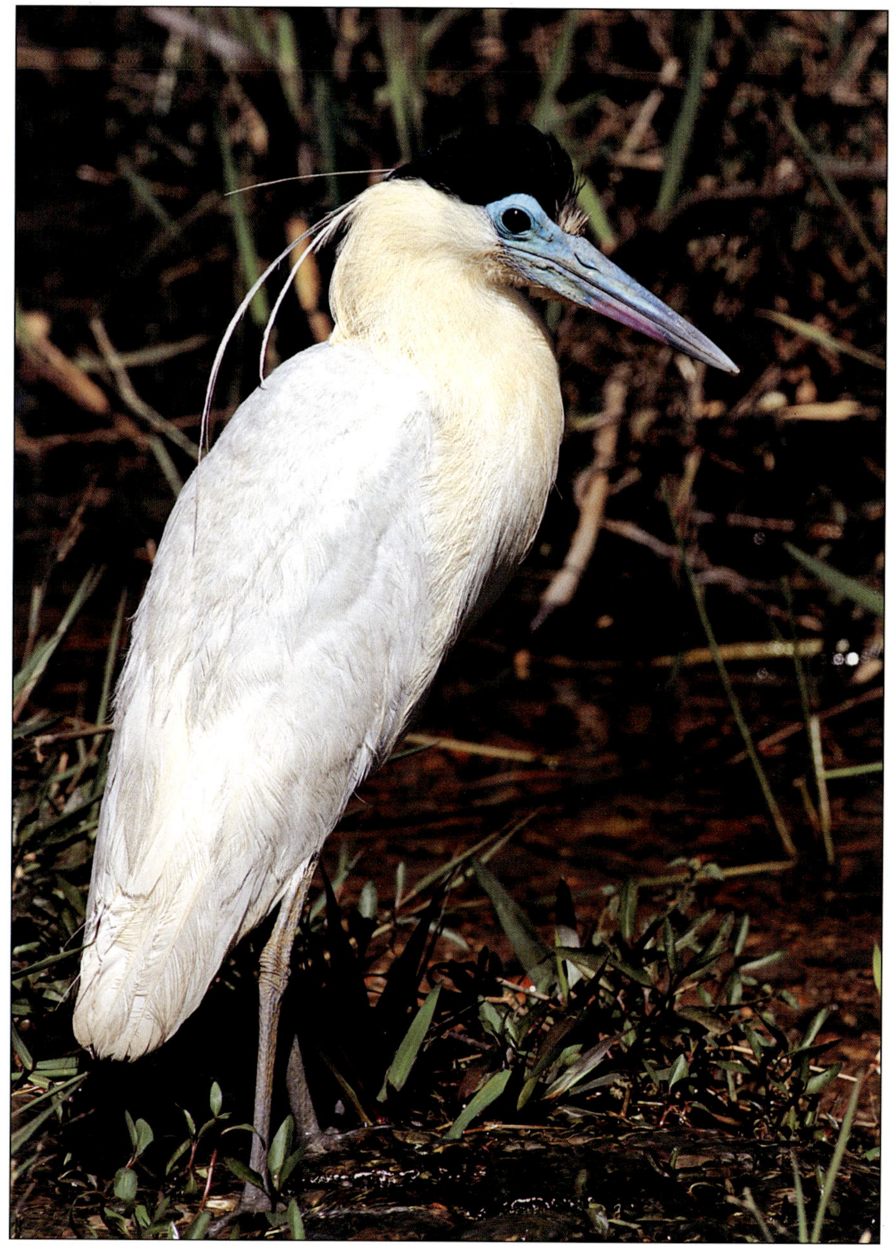

Capped Heron
Pilherodius pileatus

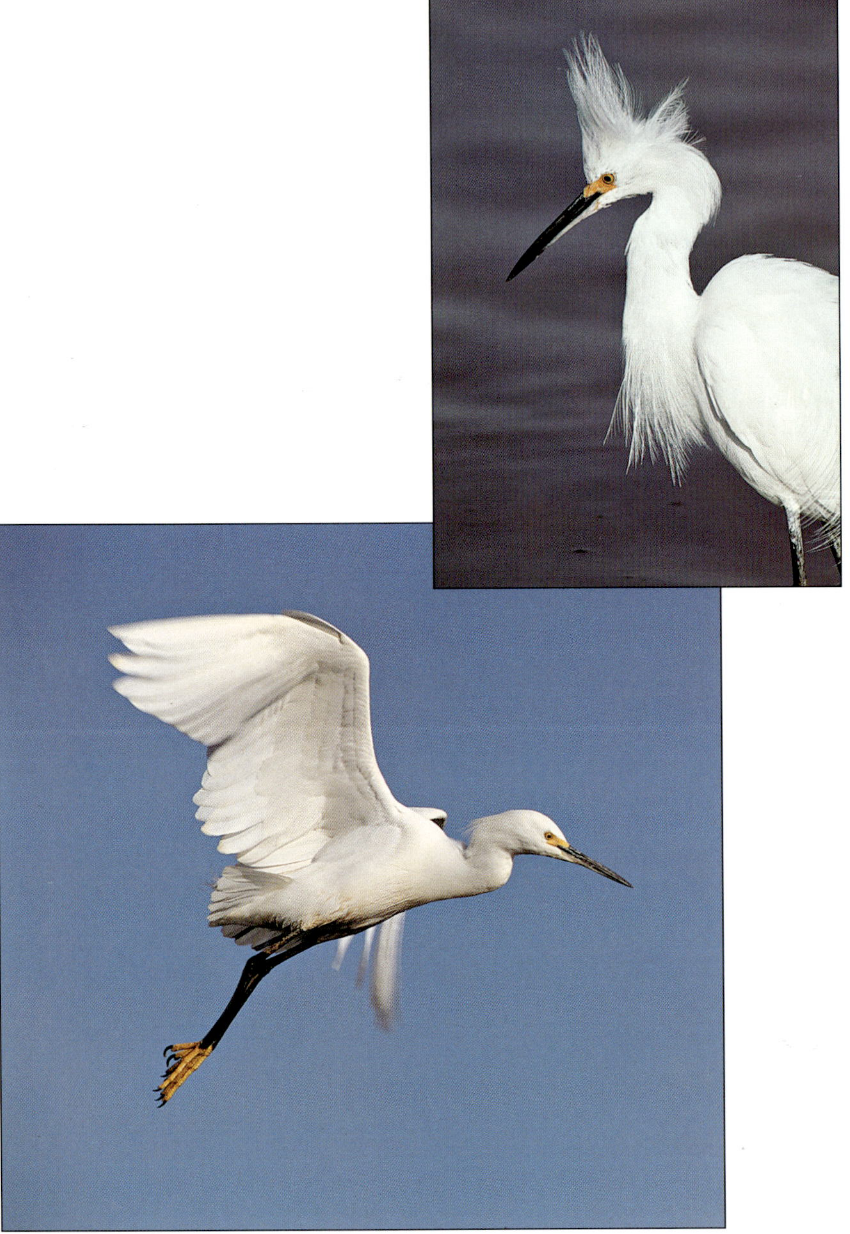

Snowy Egret
Egretta thula

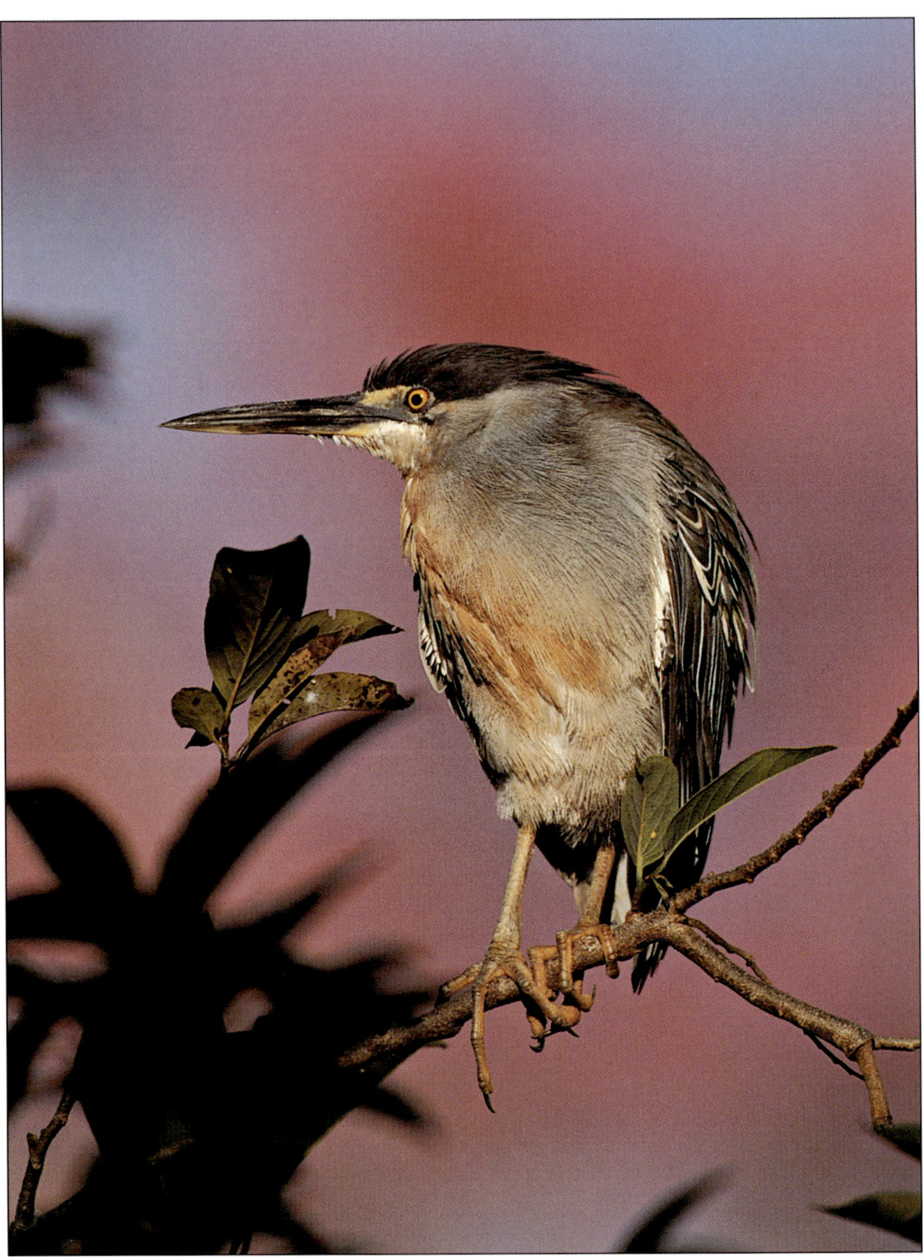

Striated Heron
Butorides striatus

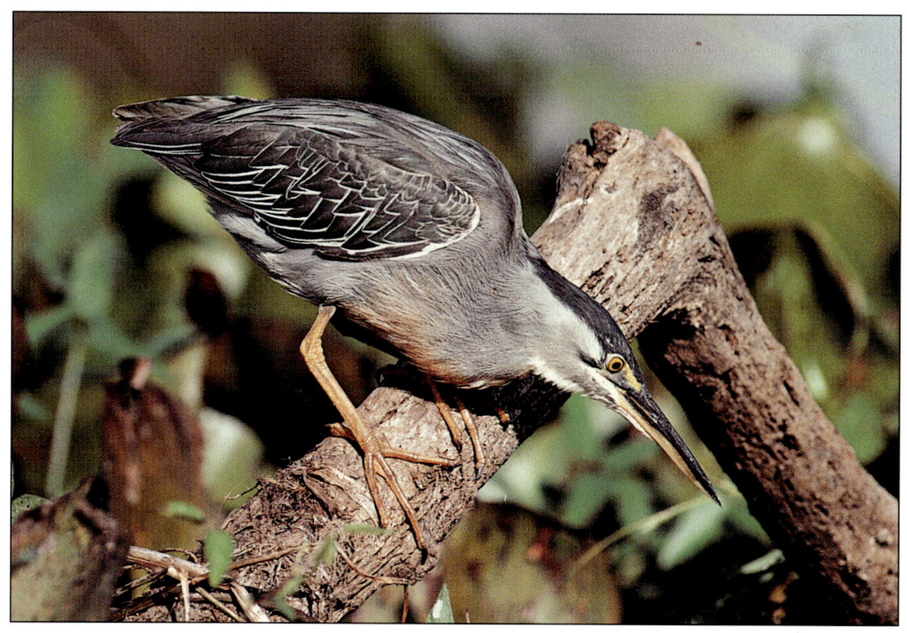

Striated Heron
Butorides striatus

A prime example for showing the virtue of patience is a Striated Heron. When stalking prey, this bird becomes so focused it will hold motionless for several minutes as it studies every move of some unsuspecting minnow. When an advantage is felt, a quick thrust of its head, neck and body usually results in a grasping catch. With a series of quick head flicks the minnow is worked up to the mouth where it is swallowed whole and head first. Since the biggest part of this bird's day is spent trying to catch something to eat, it often goes unnoticed while in the stalking position. Carefully scanning the water's edge will increase the chance of viewing this bird in action.

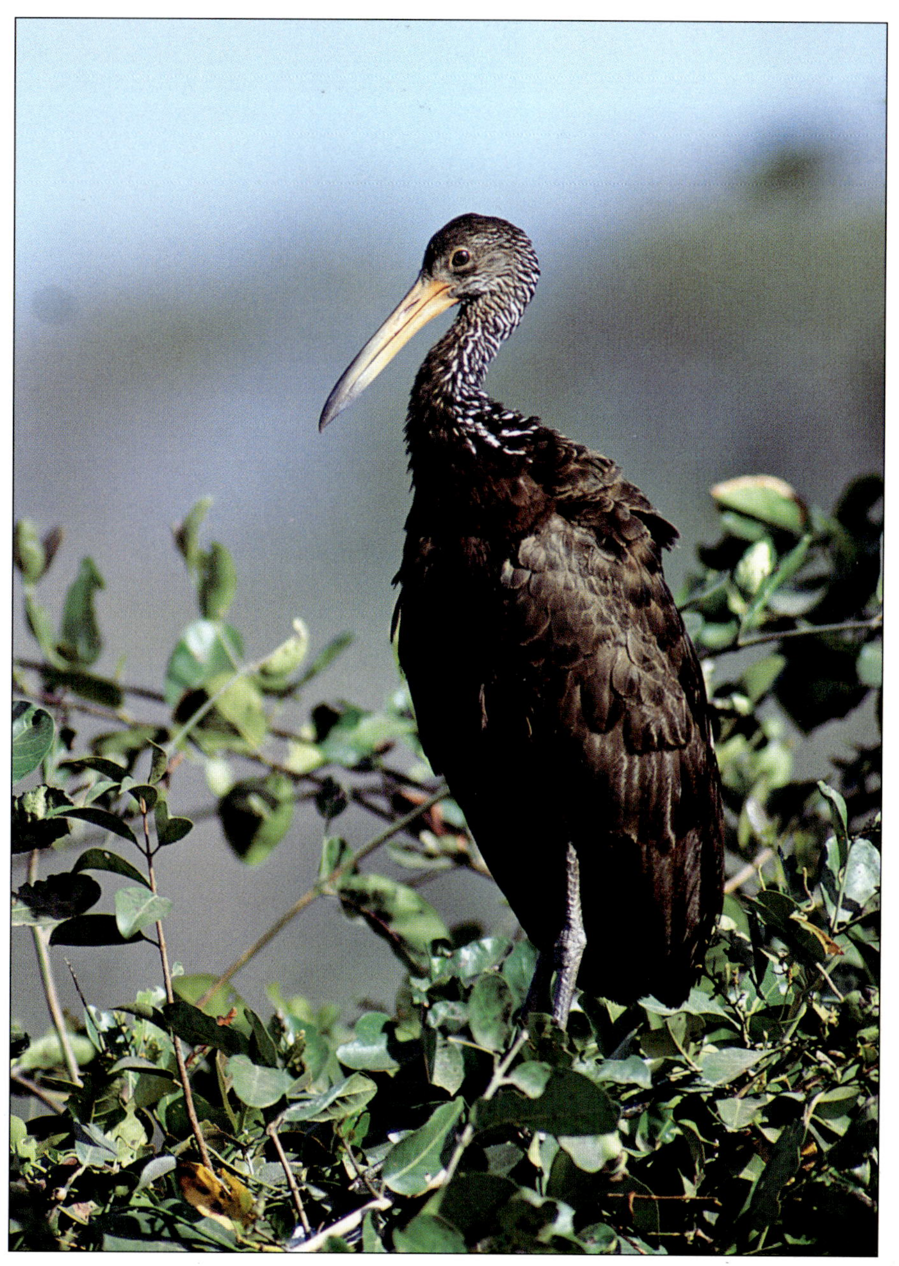

Limpkin
Aramus guarauna

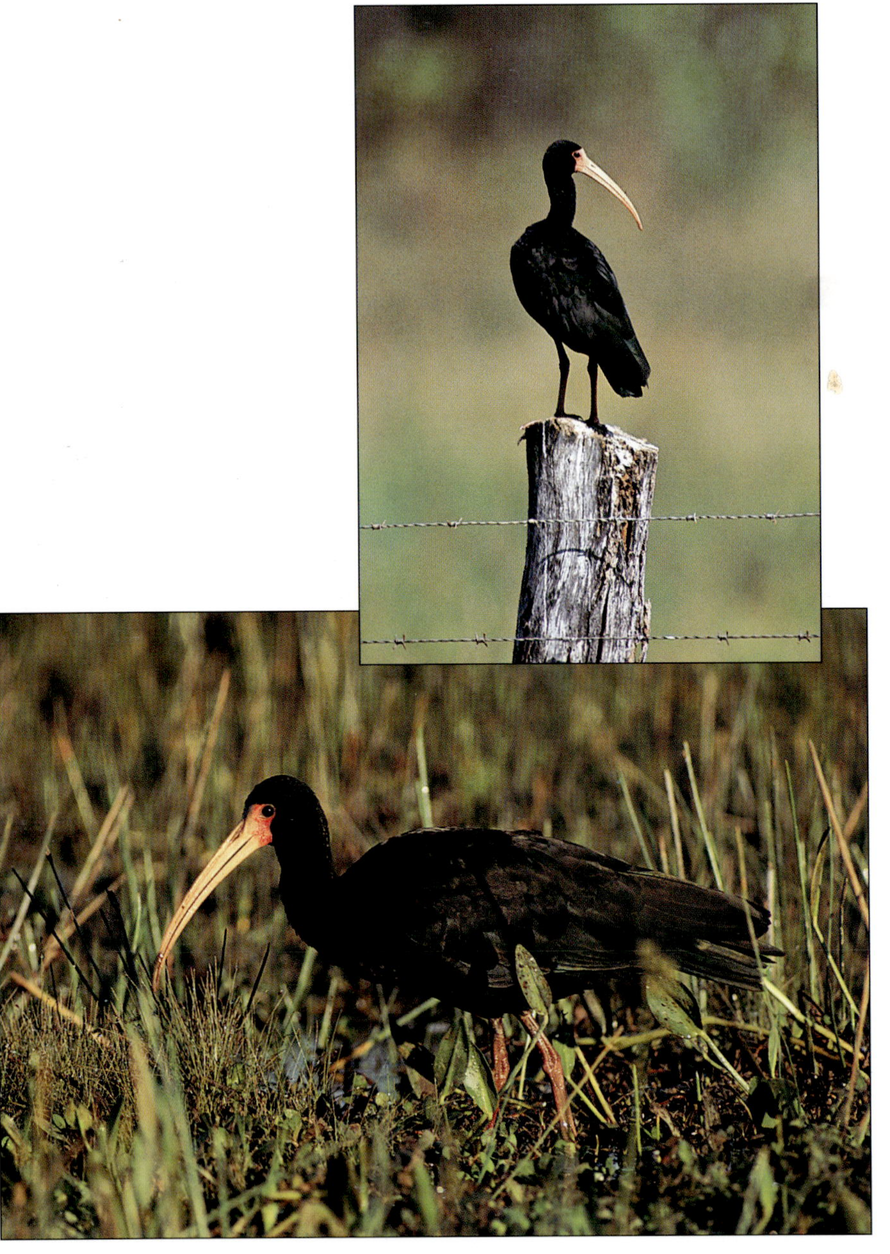

Bare-faced Ibis
Phimosus infuscatus

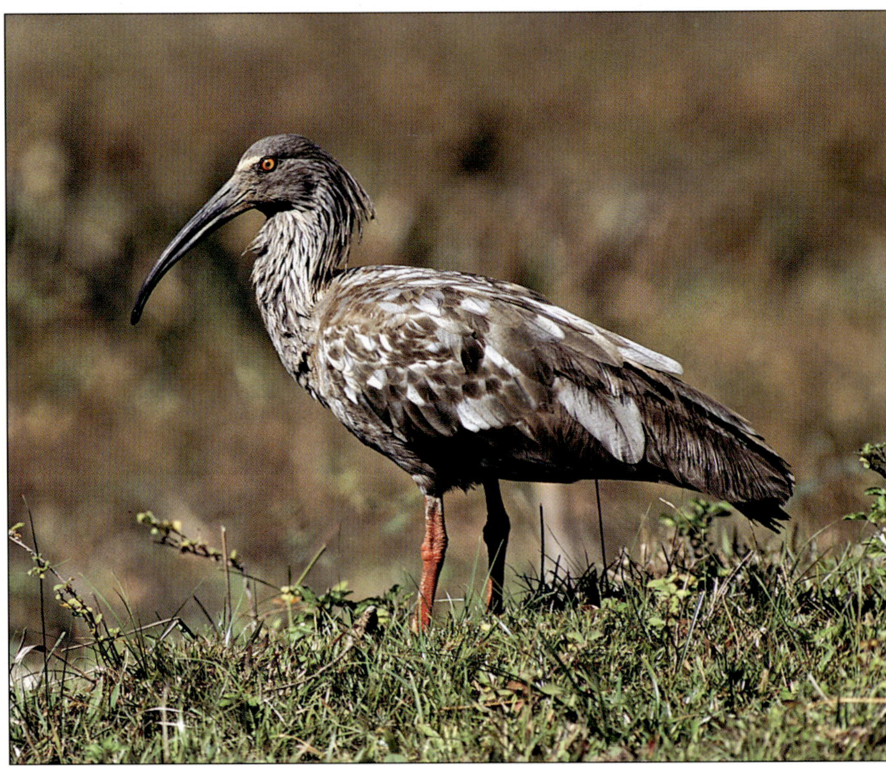

Plumbeous Ibis
Harpiprion caerulescens

Thc ibis as a group, except for one member, possess a long, down-curved, pointed bill. This bill allows them to probe mud for a variety of worms or mollusks. The tip of the bill is highly sensitive, enabling it to detect movement as it abruptly invades eight to ten inches of mud. The one exception to the group is the Roseate Spoonbill. Its bill is flattened at the ends similar to a spoon. Sweeping its head from side to side with this bill open, the spoonbill quickly snatches up any small fauna. Since the Pantanal is a huge flood plain, the ibis are well represented with five members. Except for the green ibis, most are readily viewed.

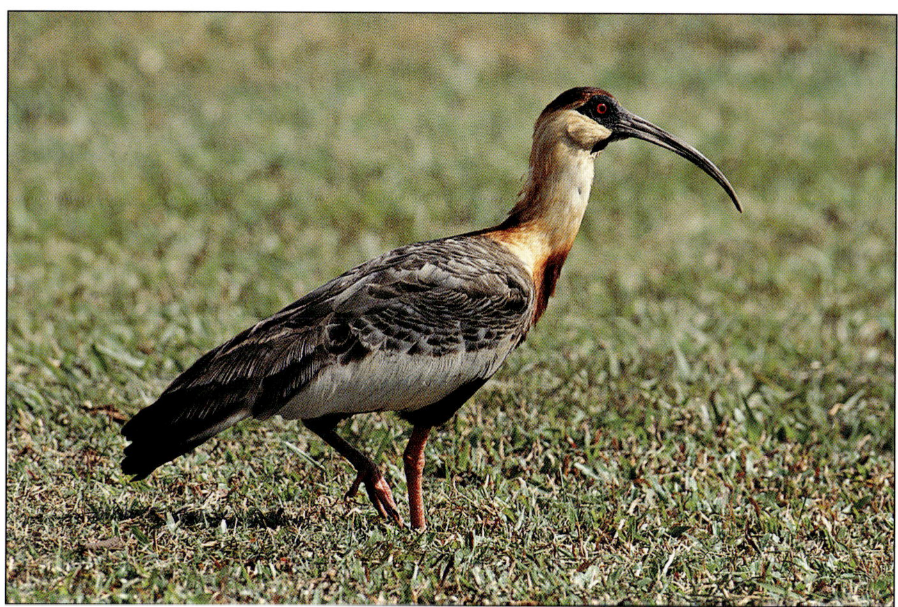

Buff-necked Ibis
Theristicus caudatus

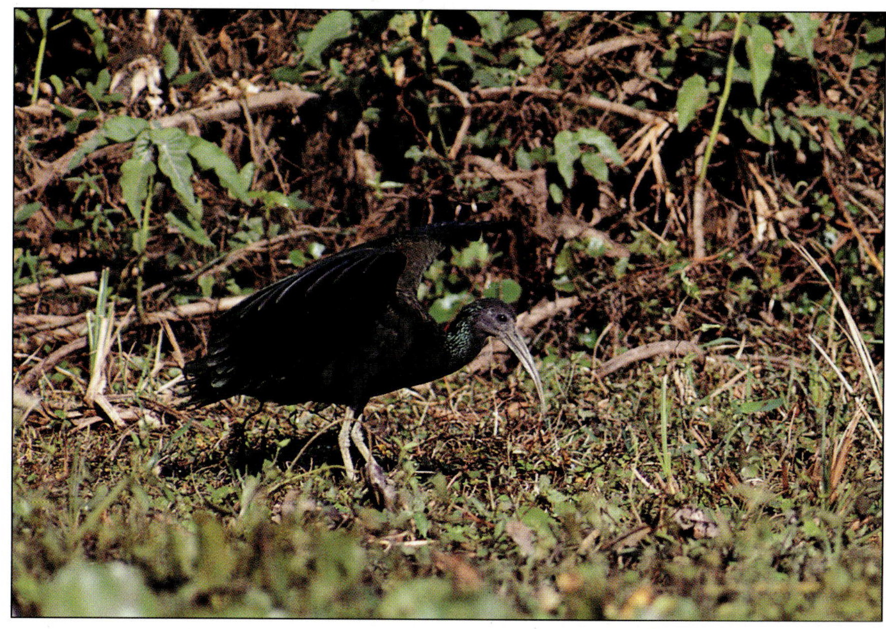

Green Ibis
Mesembrinibis cayennensis

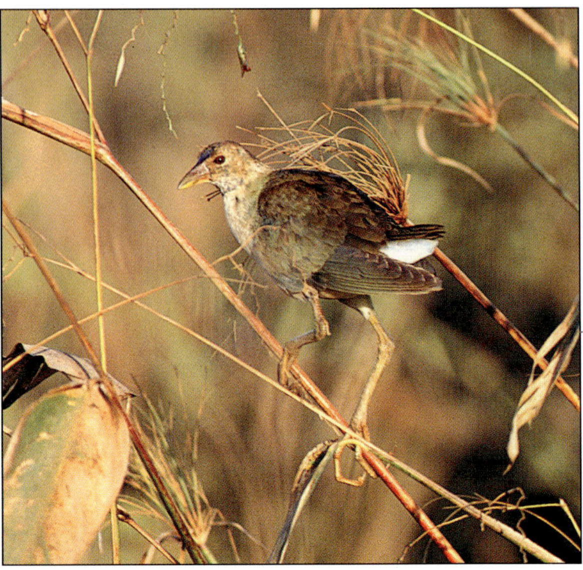

Azure Gallinule
Porphyrula flavirostris

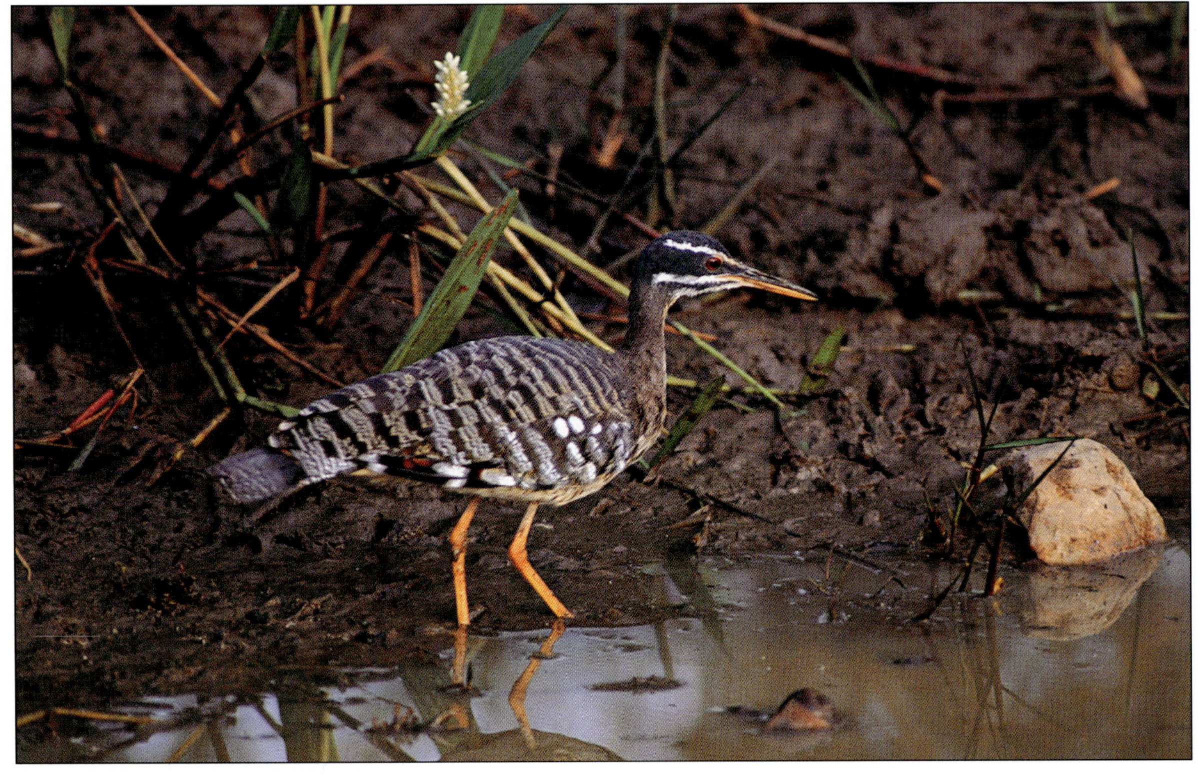

Sunbittern
Eurypyga helias

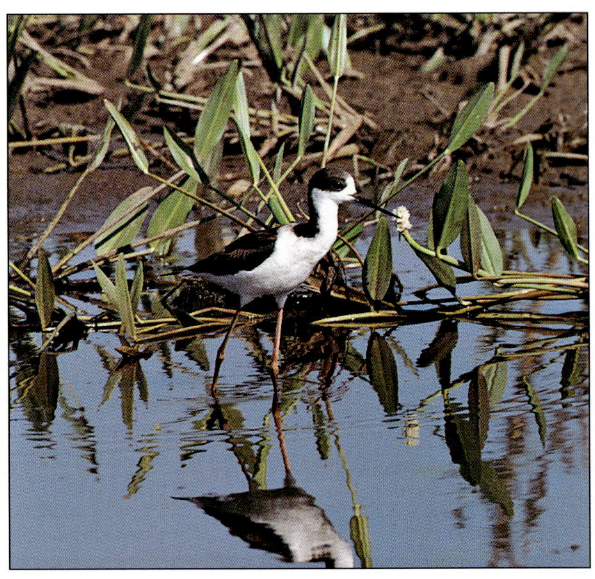

Common Stilt
Himantopus himantopus

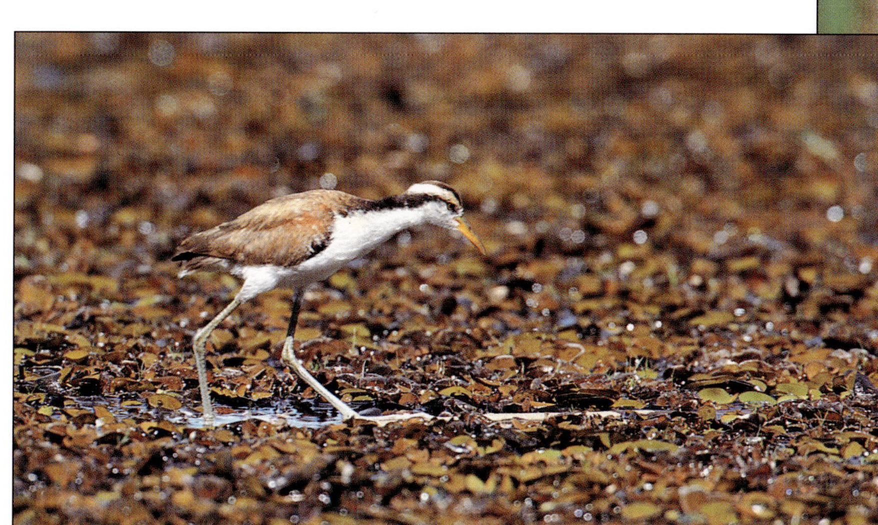

Wattled Jacana Fledgling

Jacanas belong to a group of birds that exhibit an unusual physical characteristic of their feet. Their toes are excessively elongated. When the foot is expanded, its surface area is greatly increased. This allows the bird to walk across the sparest of vegetation, investigating the surfaces of aquatic plants for adult insects and emerging larva. Most water-born insects spend much of their life beneath the surface, developing as larva or nymphs. An important stage of their life is to leave the water as an instar or nymph, shed their exoskeleton, and emerge as an adult insect. At this point they are vulnerable and become a valuable food source for hungry jacanas.

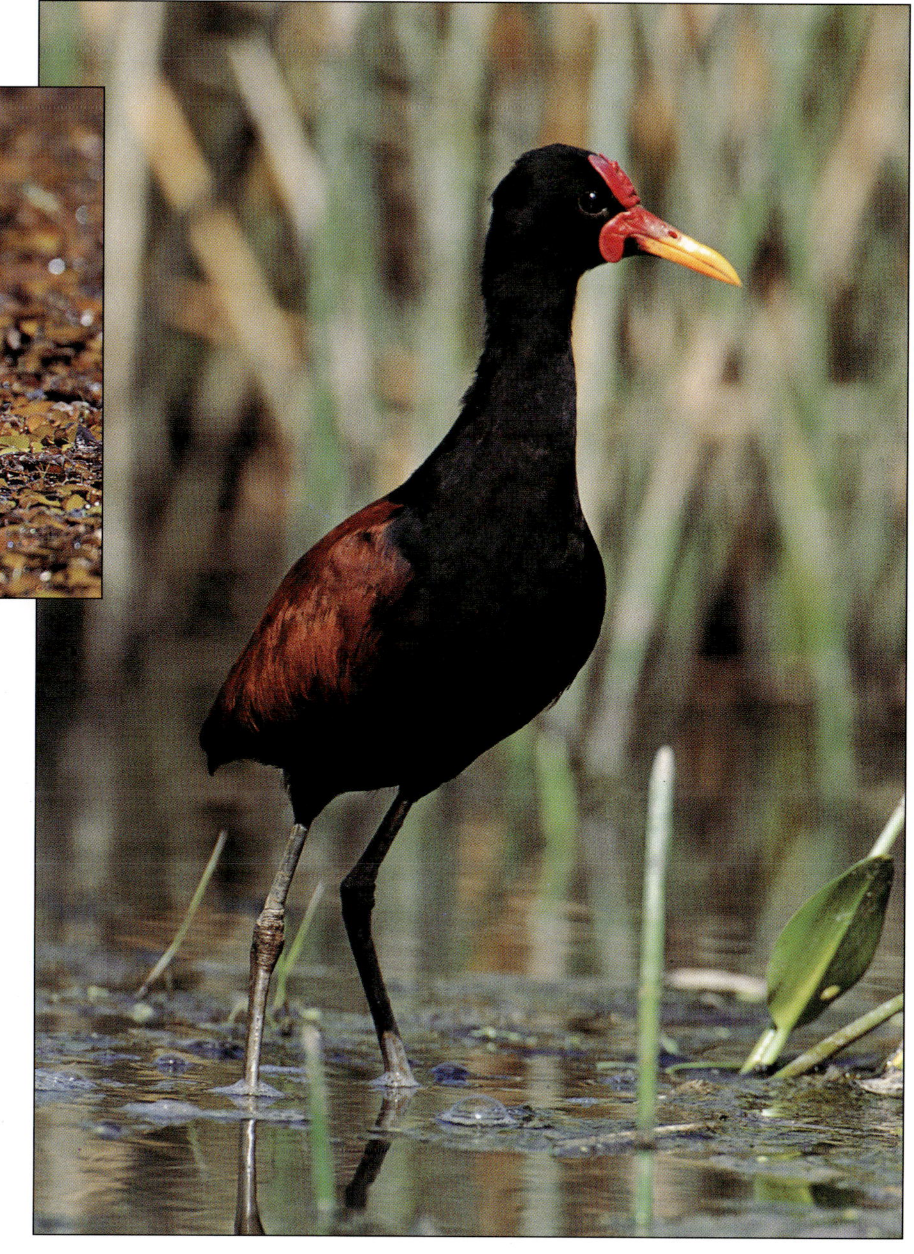

Wattled Jacana
Jacana jacana

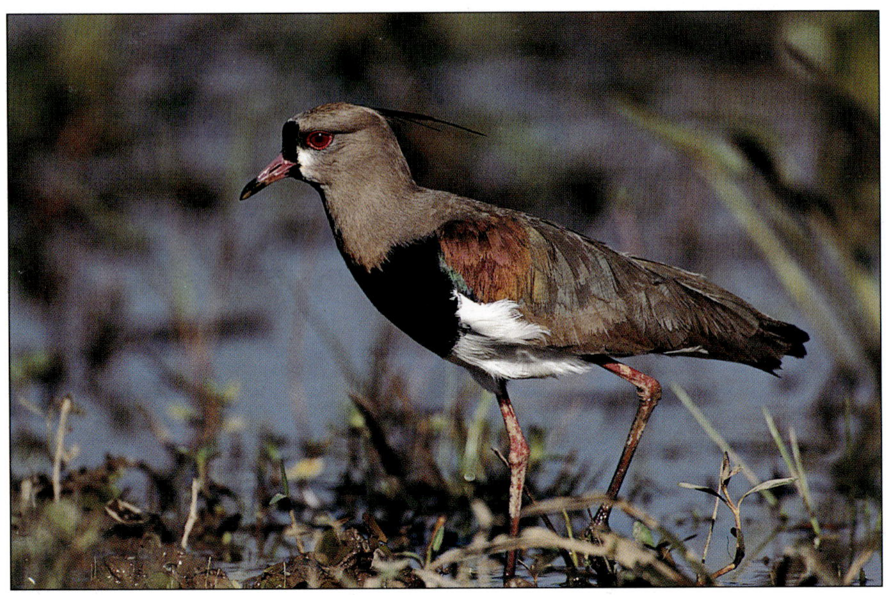

Southern Lapwing
Vanellus chilensis

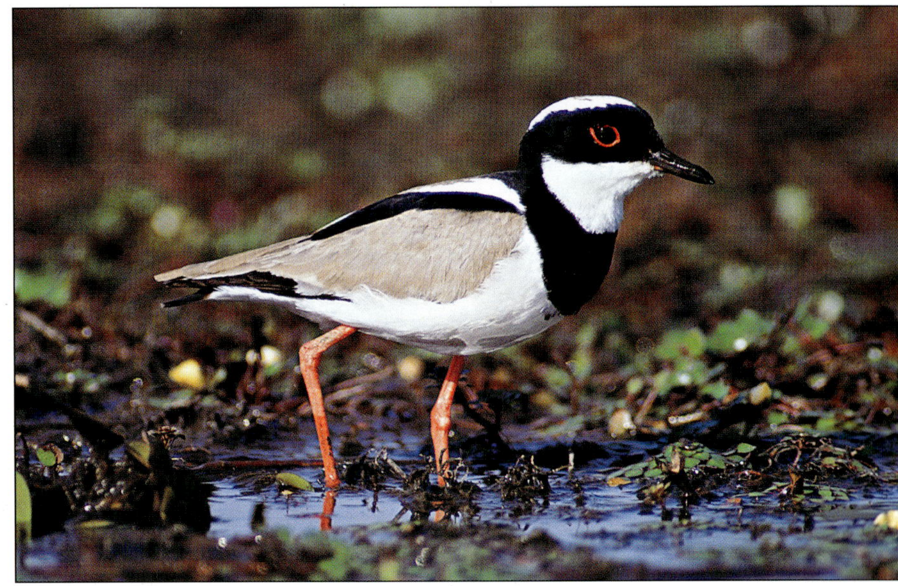

Pied Lapwing
Hoploxypterus cayanus

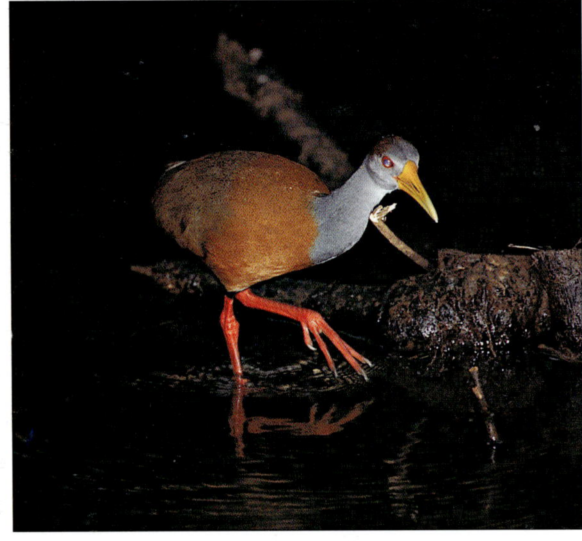

Gray-necked Wood-Rail
Armides cajanea

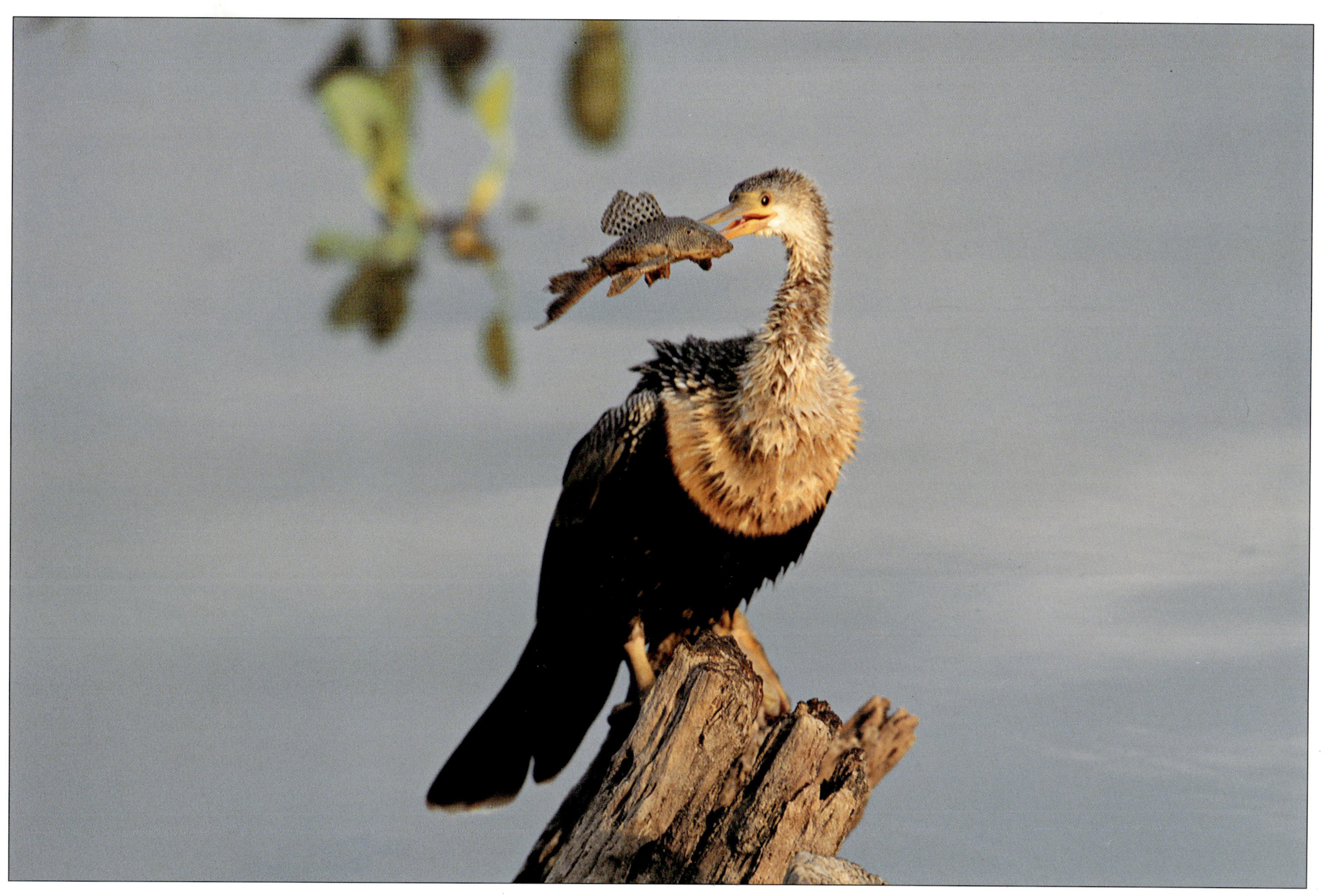

Anhinga (Freshly caught Plecoustomus)
Anhinga anhinga

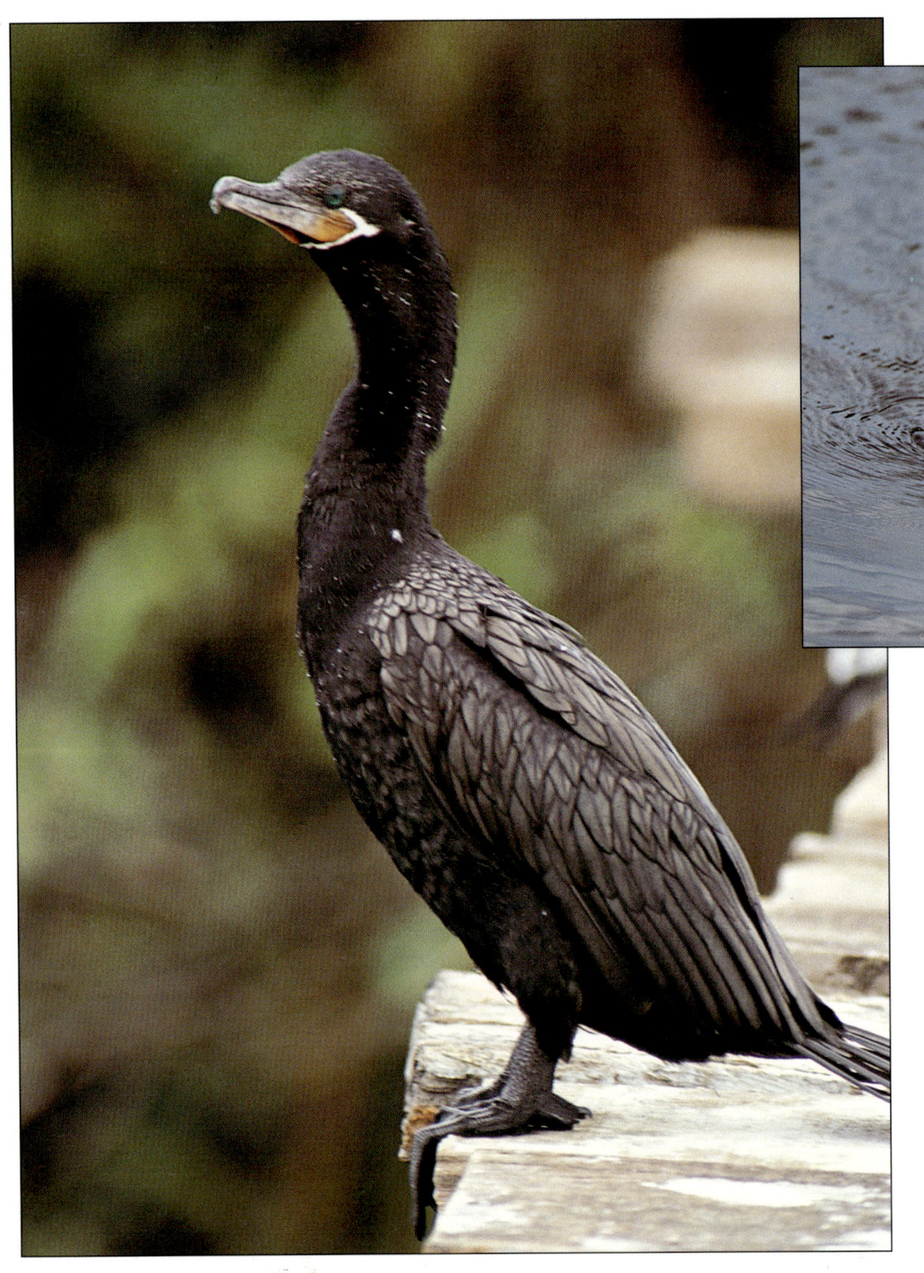

Olivaceous Cormorant
Phalacrocorax olivaceus

With a freshly caught Plecoustomus

If cormorants and/or anhingas are present, one can be sure fishing is good. These birds are extremely efficient at what they do. They actively fish, and are good at it. By compressing feathers close to the body to eliminate air, they attain neutral buoyancy and submerge easily. Propelled by huge webbed feet, these birds achieve mobility beneath the surface which is akin to flying. With necks coiled back, their heads strike out with deadly accuracy when they come upon an unsuspecting fish. The Anhinga impales its prey with a sharp dagger-like bill while the cormorant just snatches it up with its bill. When their feeding period is over, both species must perch usually facing the sun with wings outstretched, to dry themselves.

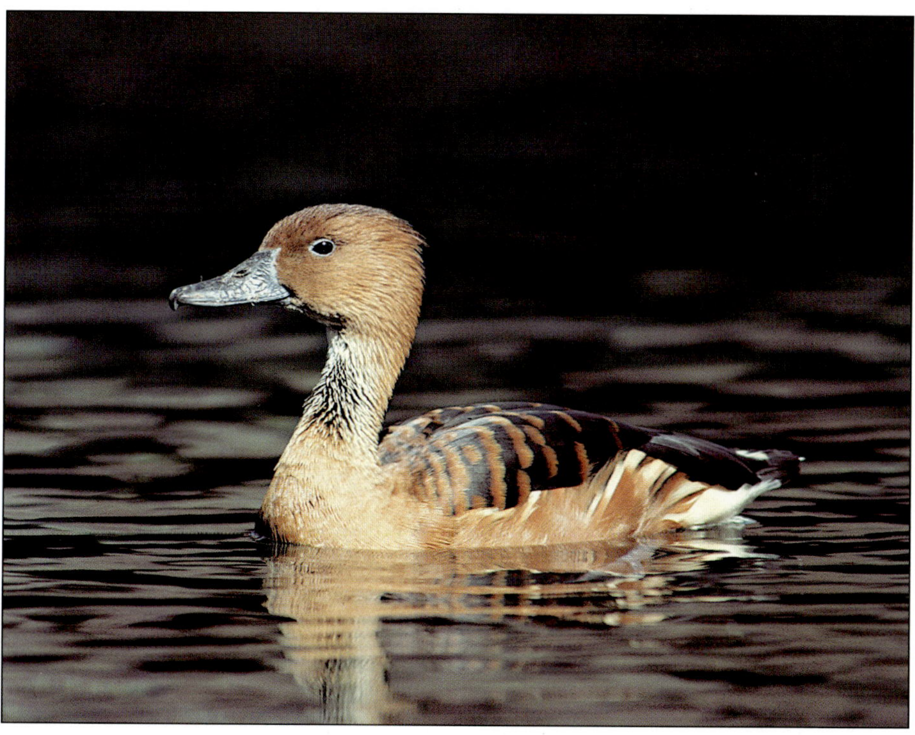

Fulvous Whistling Duck
Dendrocygna bicolor

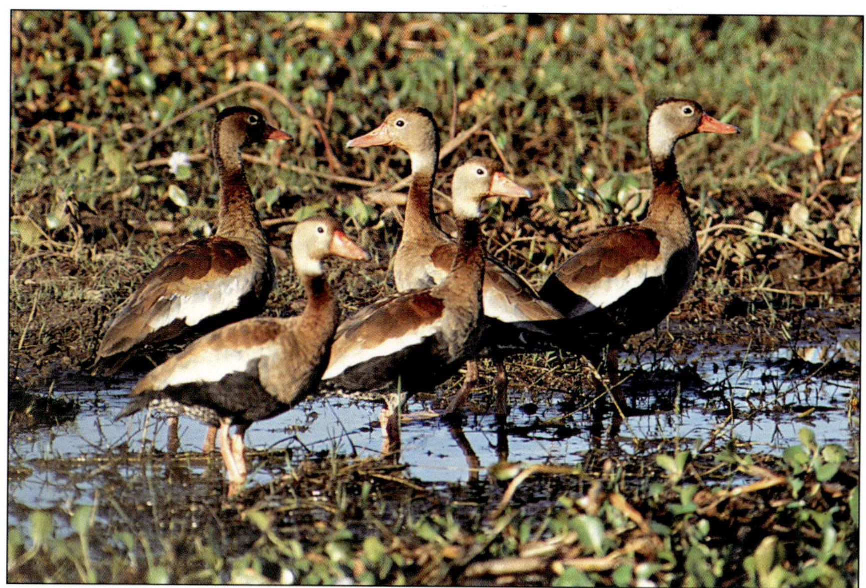

Black-bellied Whistling Duck
Dendrocygna autumnalis

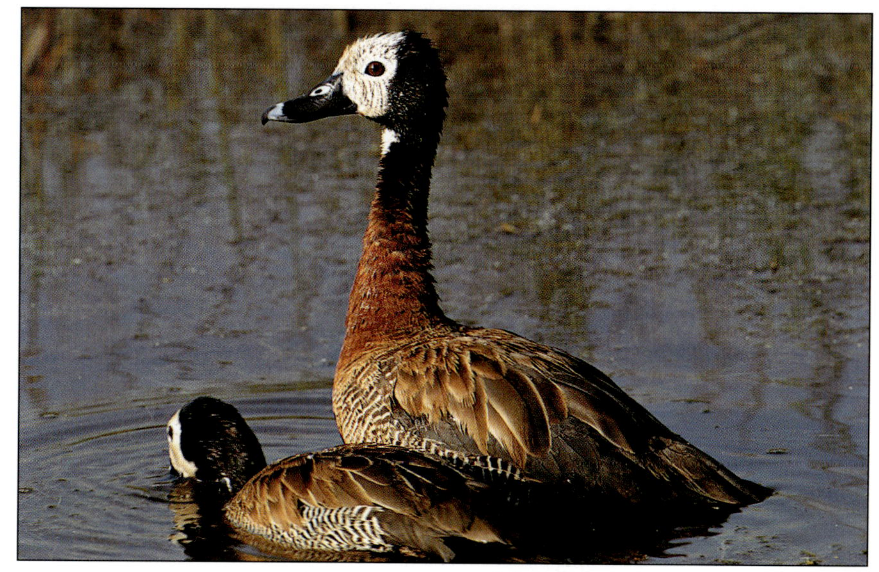

White-faced Whistling Duck
Dendrocygna viduata

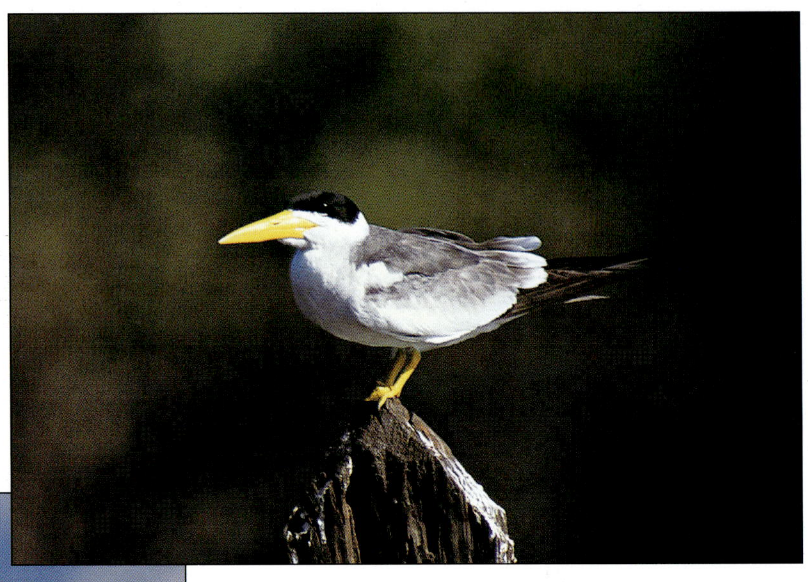

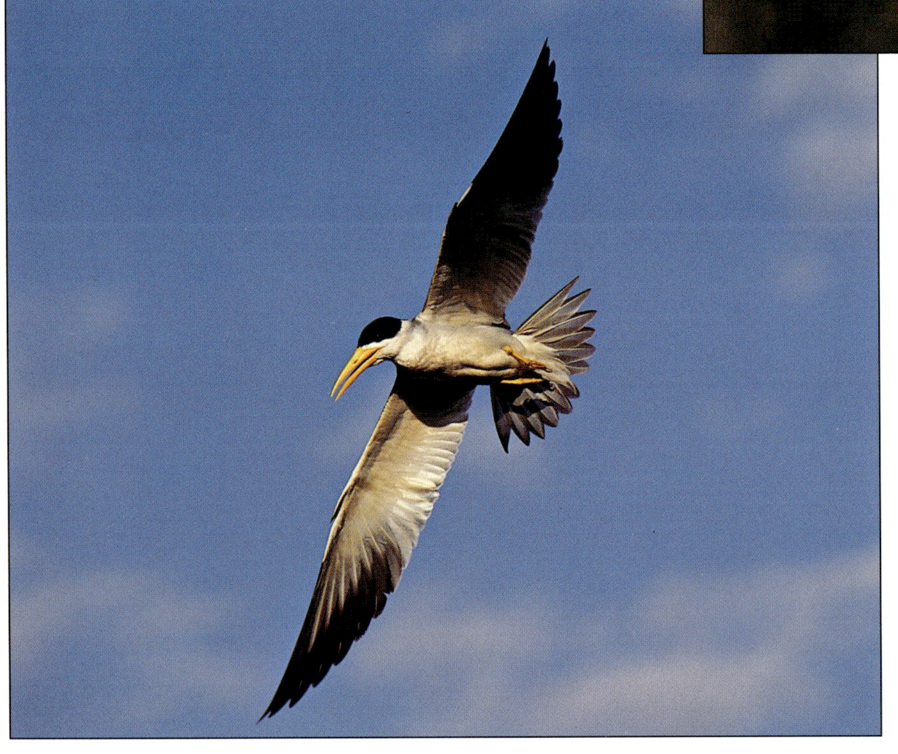

Large-billed Tern
Phaetusa simplex

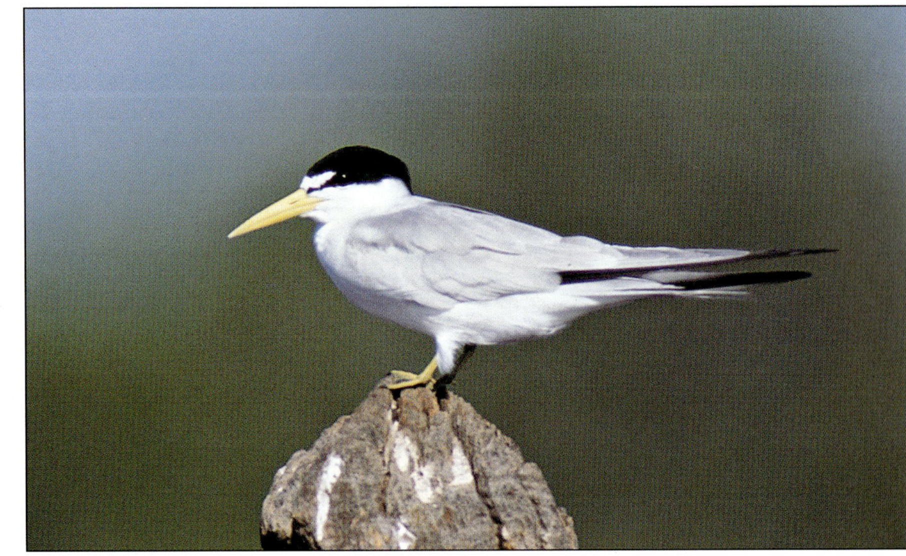

Yellow-billed Tern
Sterna superciliaris

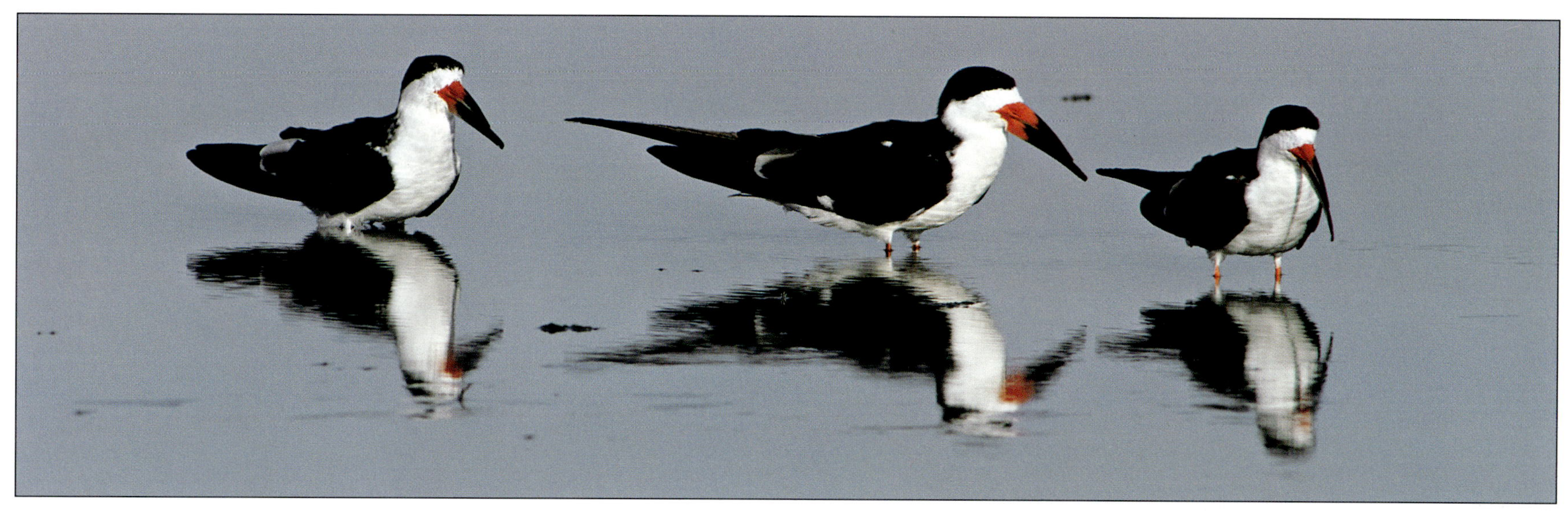

No other avian species in the world fishes for its meals like the skimmers. Flying low over water's surface, this bird slices through the water with its lower bill. Any small fish basking below the surface are quickly snatched up and swallowed. One often wonders how this technique could be so efficient. One reason is that microscopic plankton collect near the surface to utilize the sun's energy. This attracts the smaller organisms that feed upon this concentration of plankton. These smaller organisms in turn become food for the smaller fish.

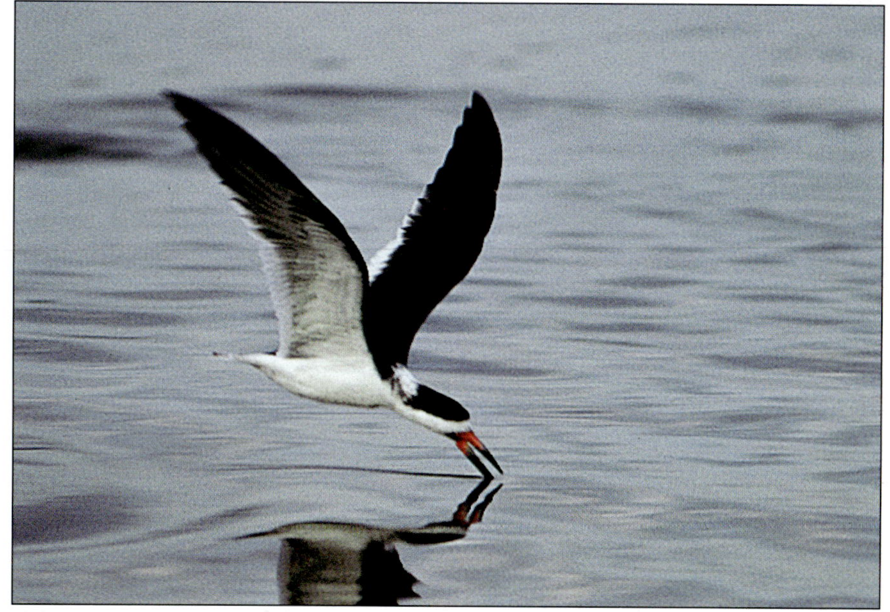

Black Skimmer
Rynchops niger

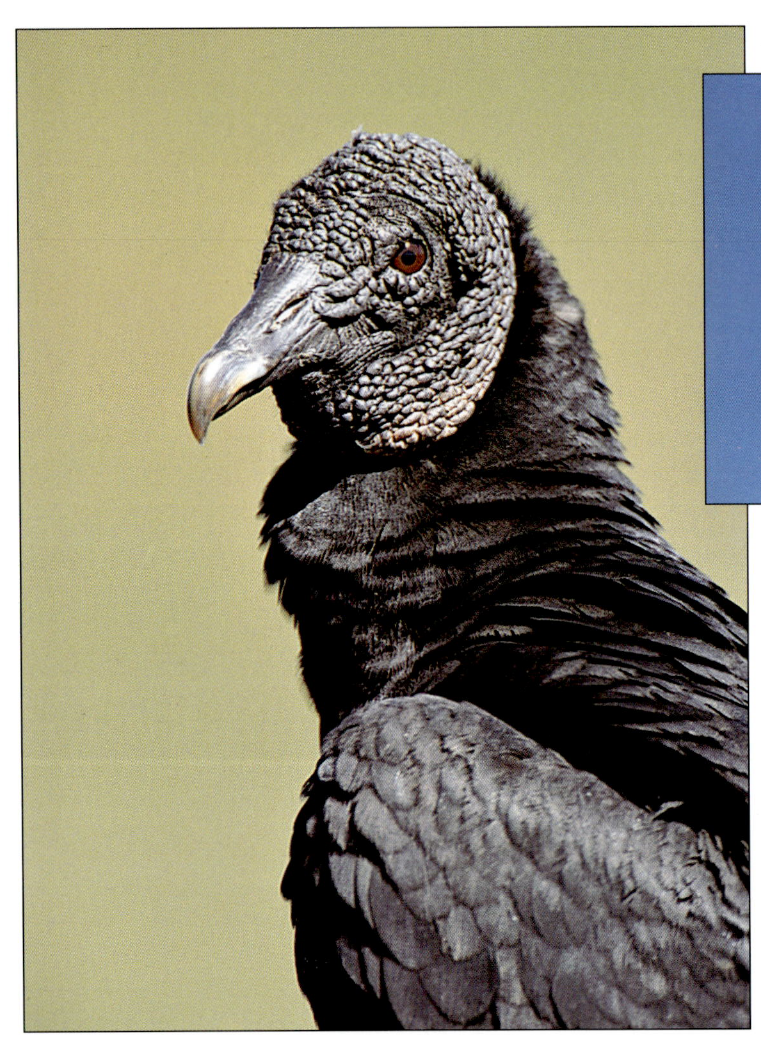

Black Vulture
Coragyps atratis

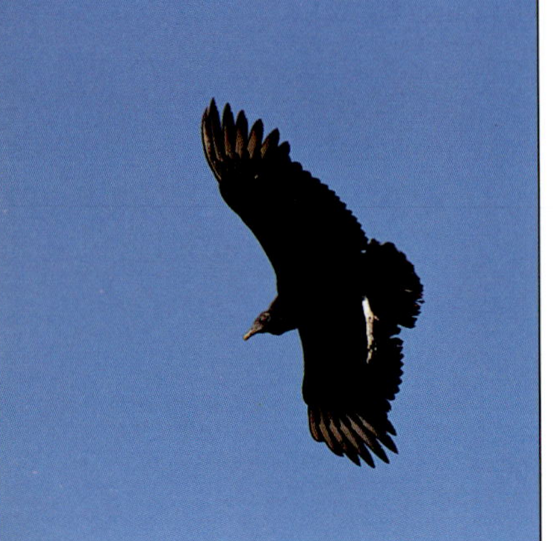

Turkey Vulture
Cathartes aura

Lesser Yellow-headed Vulture
Cathartes burrovianus

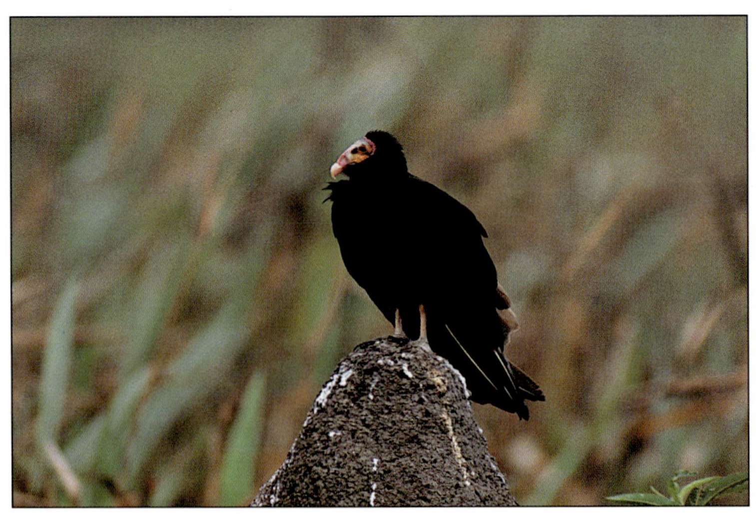

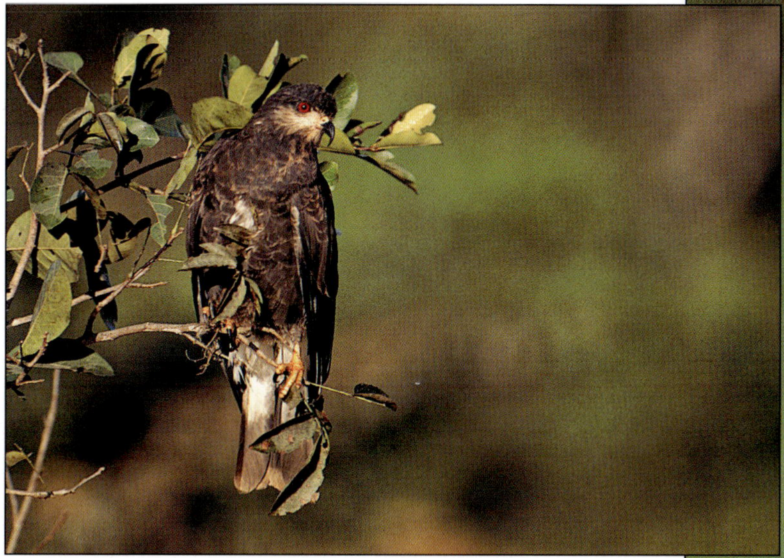

Immature Snail Kite

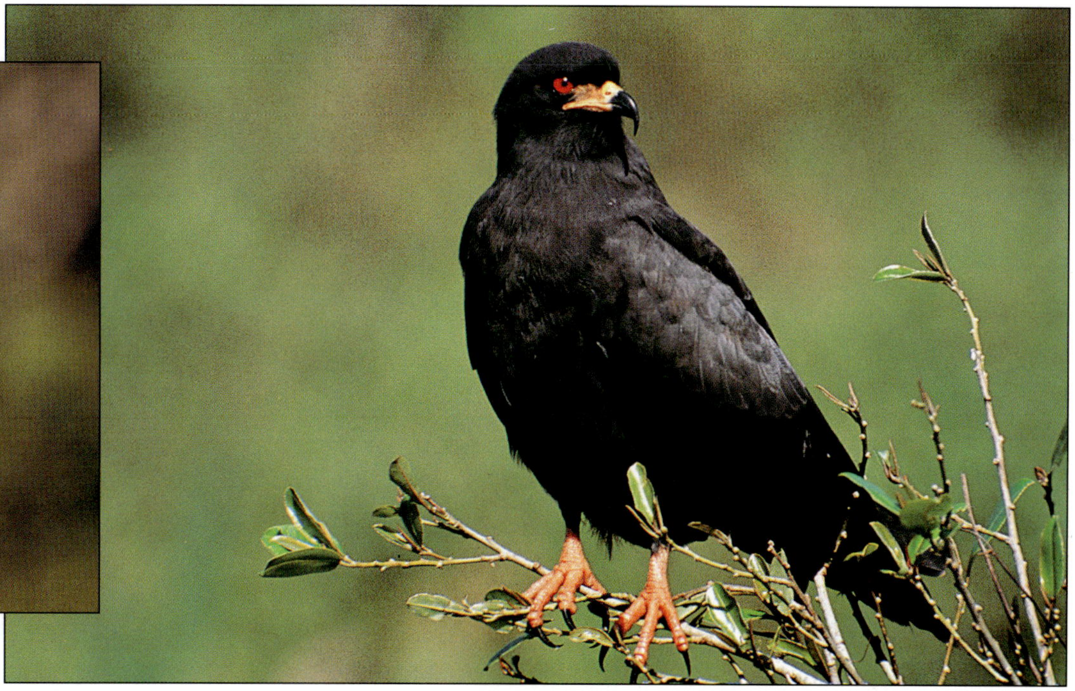

Snail Kite
Rostrhamus sociabilis

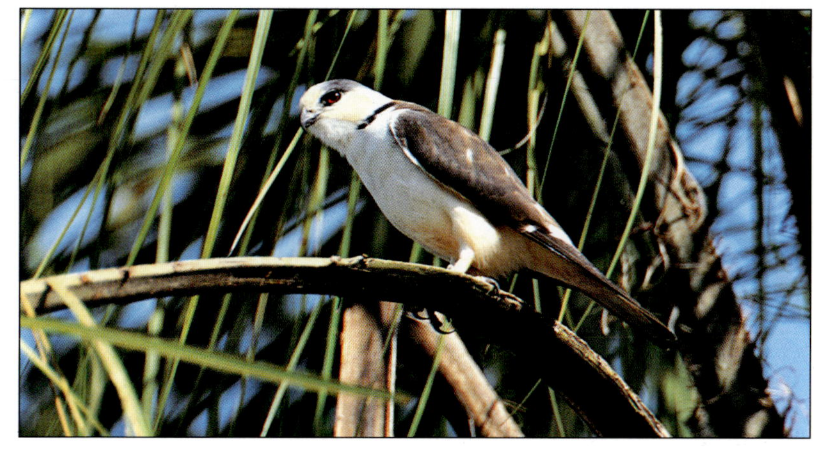

Pearl Kite
Gampsonyx swainsonii

The Snail Kite is one of the most prominent hawks in the Pantanal. It is not uncommon to view one every few hundred meters while driving the Transpantaneria. Adult males are slatey-blue with bright red eyes. Females and immature males are more difficult to identify. Brown on top and down the back, they may also have some streaking down the front. All of them have a pronounced hooked tip to the upper bill. They insert this elongated tip as a tool to cut the columellar muscle of snails, their favorite food. Cutting this muscle frees the snail from its shell for easy removal. It is also thought the red of the male's eye helps filter out the blues and greens of the water, allowing the Snail Kite to see more clearly beneath the surface when hunting snails.

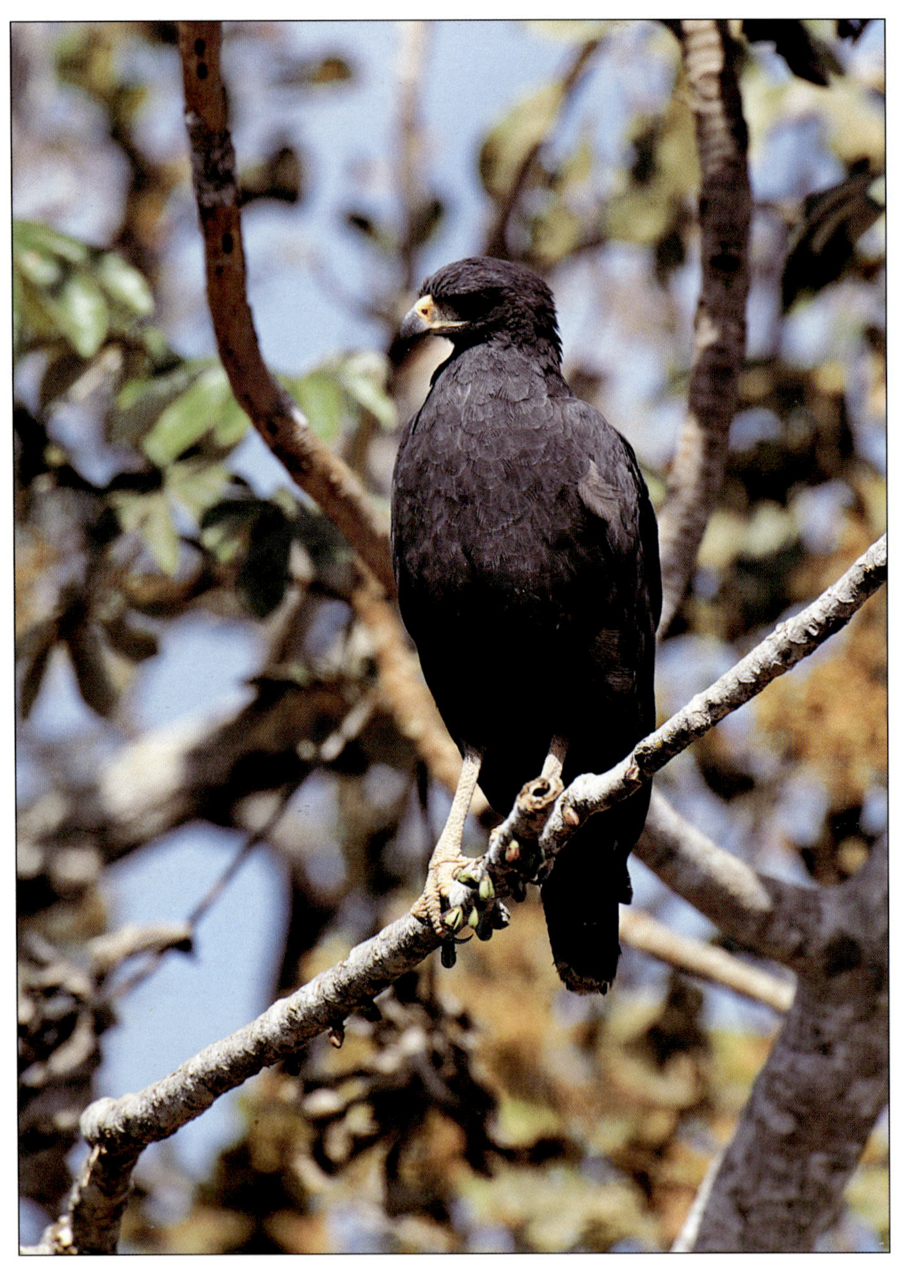

Great Black Hawk
Buteogallus urubitinga

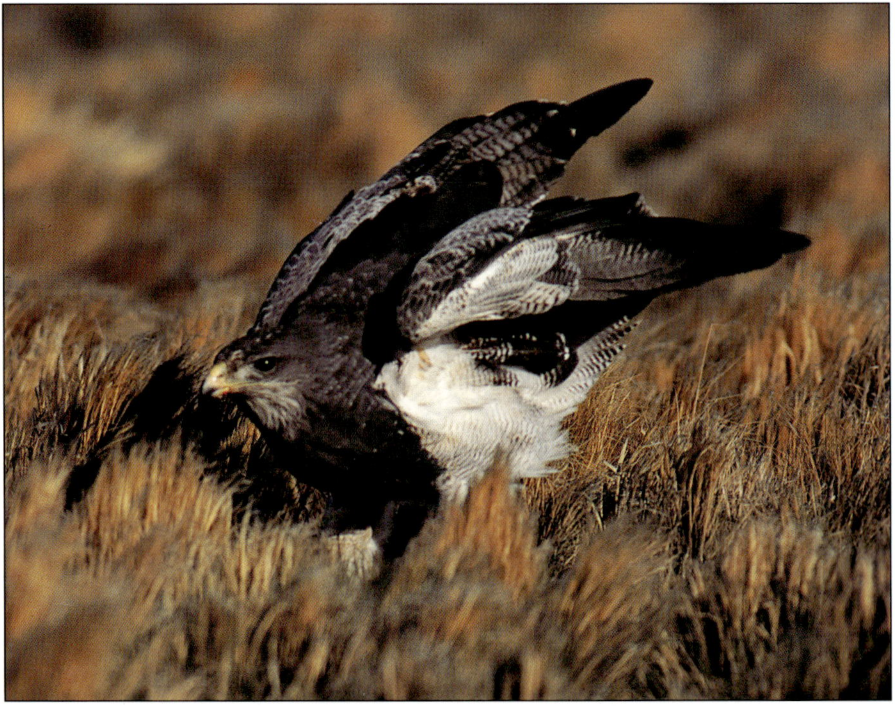

Black-chested Buzzard-Eagle
Geranoaetus melanoleucus

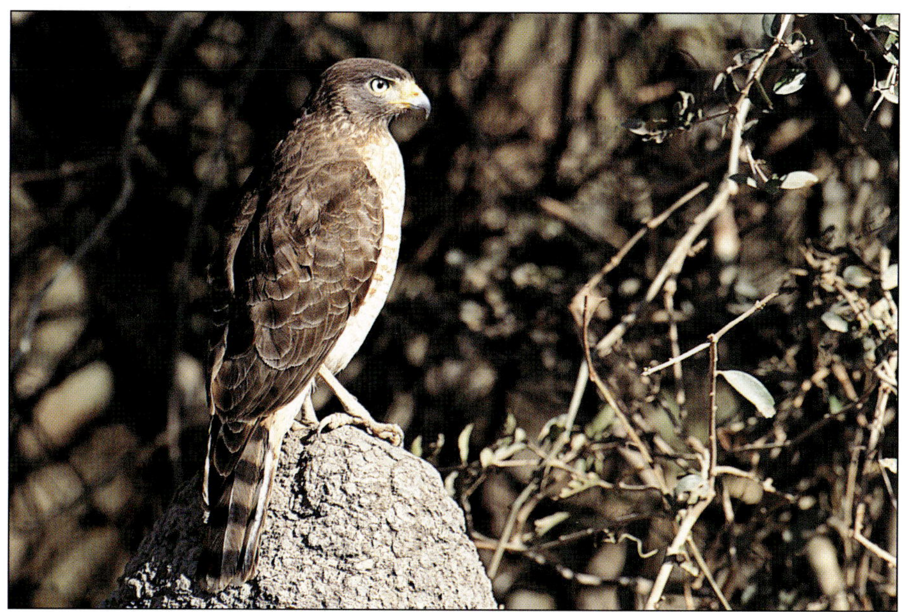

Roadside Hawk
Buteo magnirostris

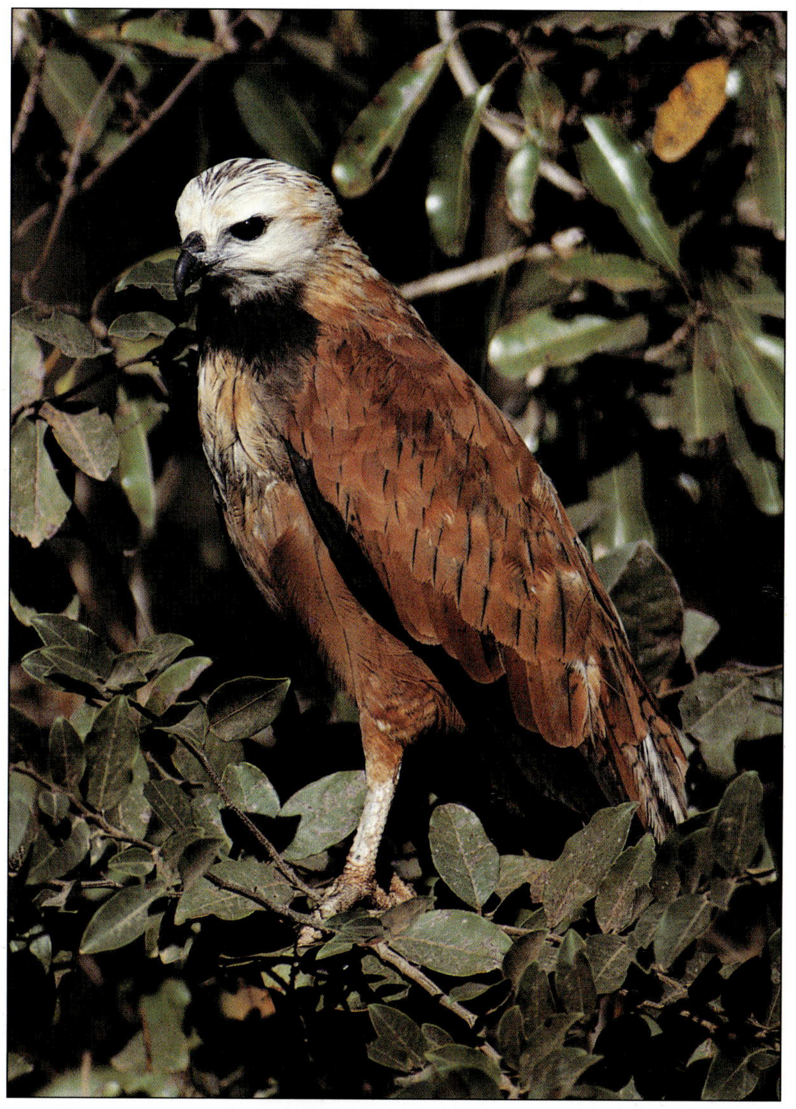

Black-collared Hawk
Busarellus nigricollis

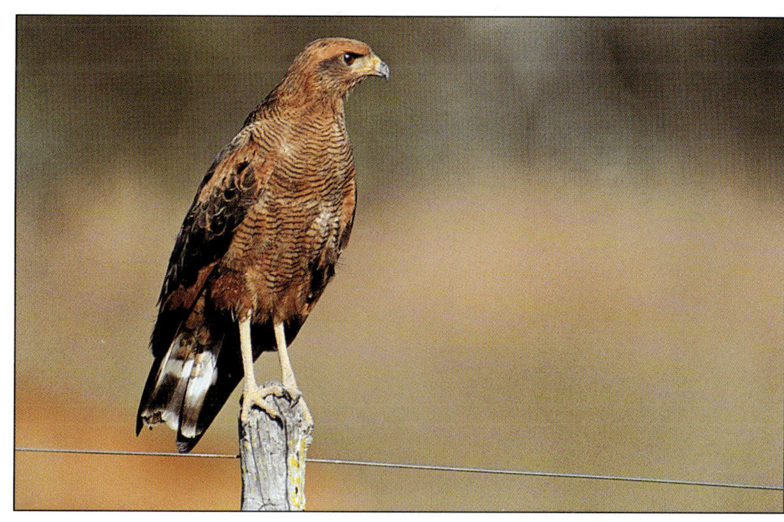

Savanna Hawk
Heterospizias meridionalis

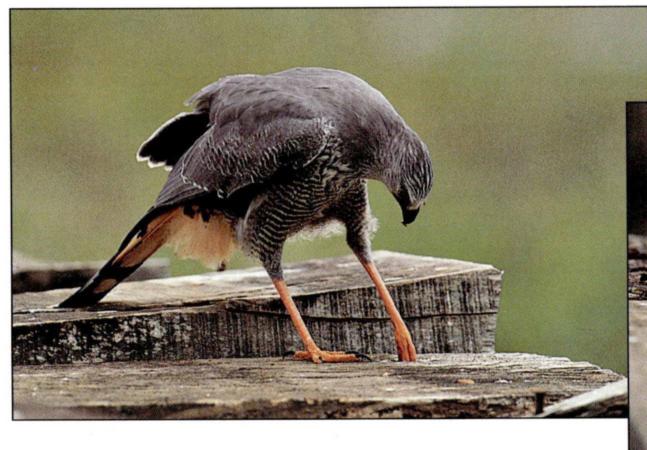

Crane Hawk
Geranospiza caerulescens

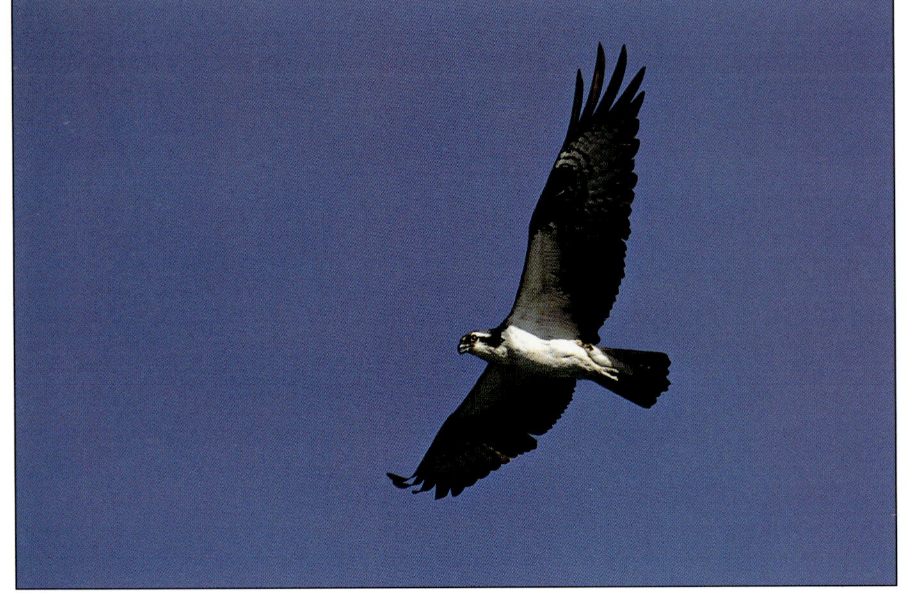

Osprey
Pandion haliaetus

The Crane Hawk receives its name from the way the bird uses its long thin legs to probe holes in trees for small lizards, insects, or even bats. One time, for close to an hour, this bird was observed investigating and probing the holes and cracks on a bridge along the Transpantaneria. After all its work, the search was unsuccessful.

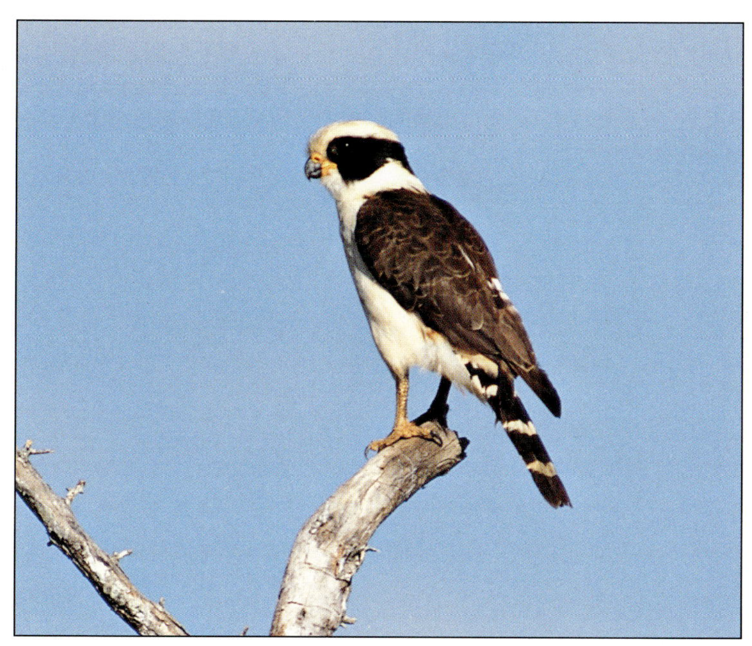

Laughing Falcon
Herpetotheres cachinnans

Often considered a scavenger, yet classified as a falcon, the Crested Caracara inhabits most of South America. In fact, it ranges north to Texas. With this expansive range, it clearly shows how adaptive they can be. Almost anything from insects, fish, baby caiman to carrion is fair game for the caracara. Much time is spent in the company of vultures which shows they are bold enough to compete with them even while away from carcasses. Many will gather near fishermen waiting for scraps left after cleaning fish. The bare reddish area of the Crested Caracara's face becomes bright yellow when this bird is excited.

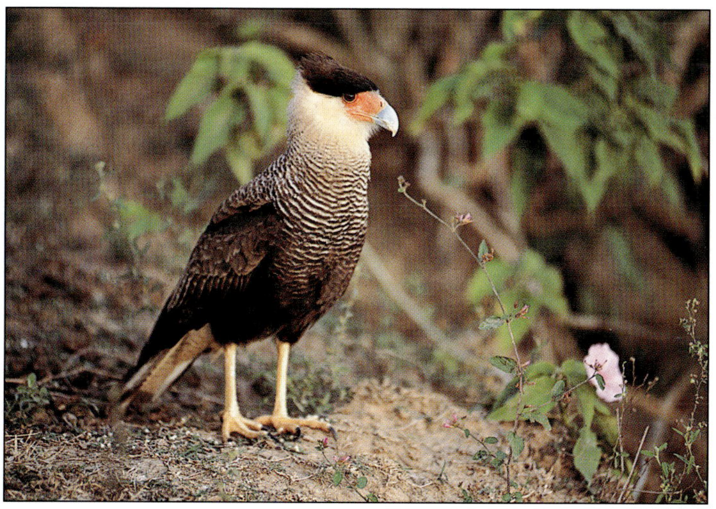

Crested Caracara
Polyborus plancus

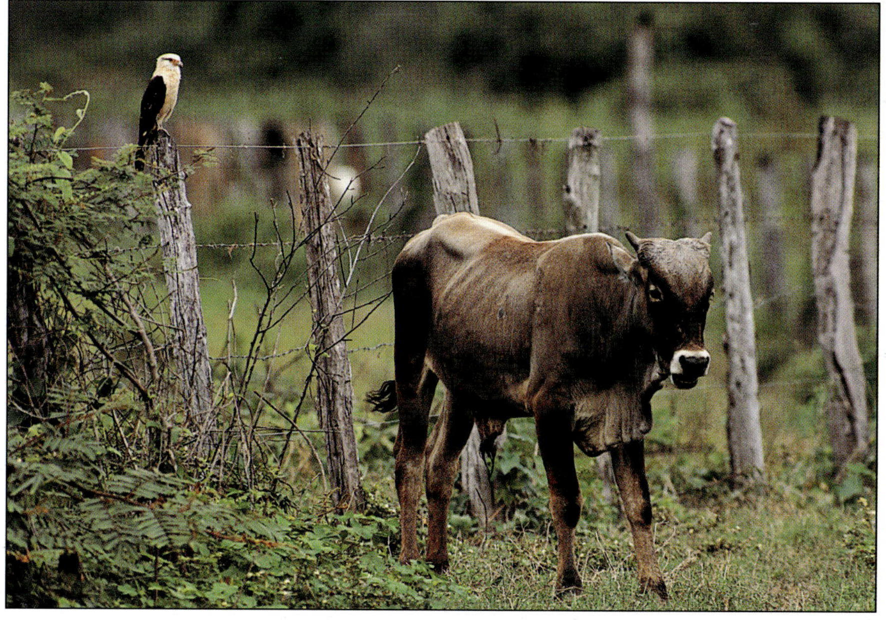

Yellow-headed Caracara
Milvago chimachima

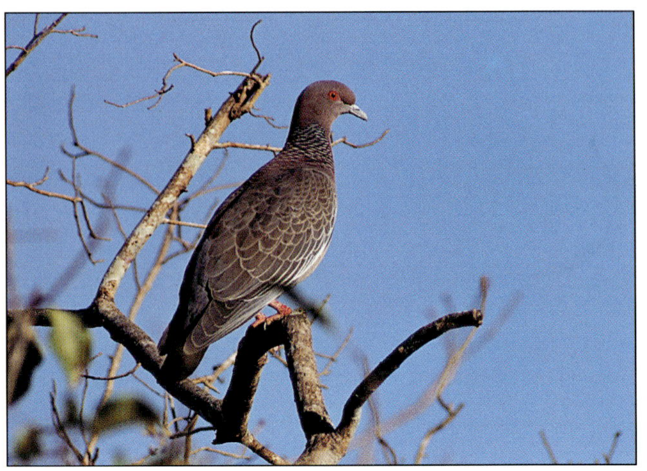

Picazuro Pigeon
Columba picazuro

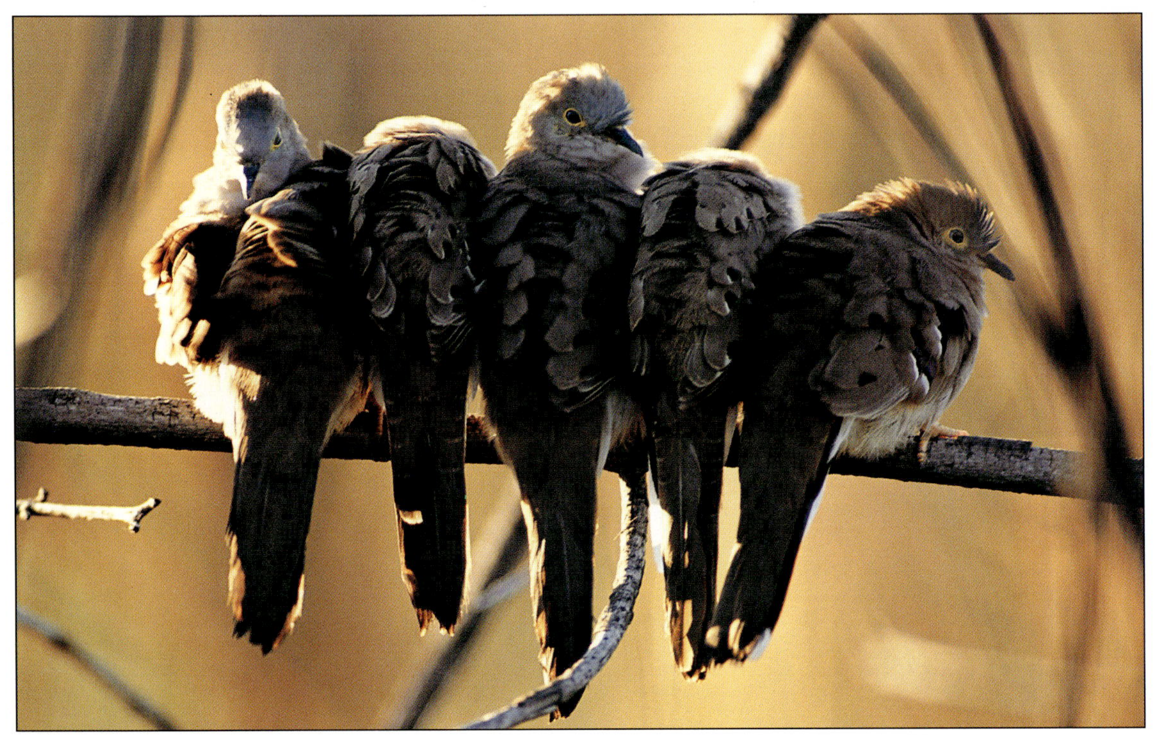

Picui Ground Dove
Coumbina picui

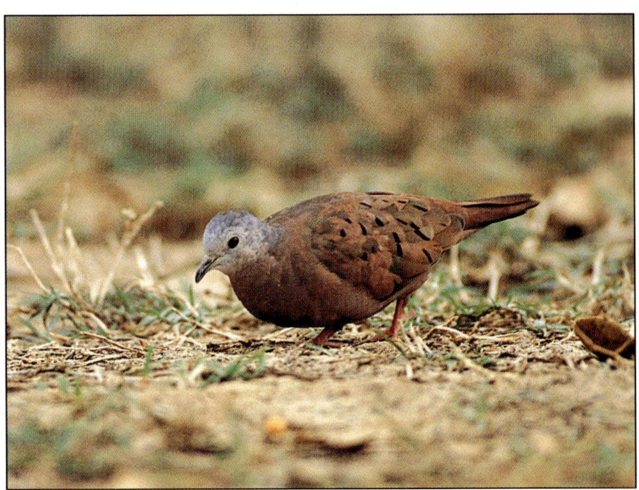

Ruddy Ground Dove
Columbina talpacoti

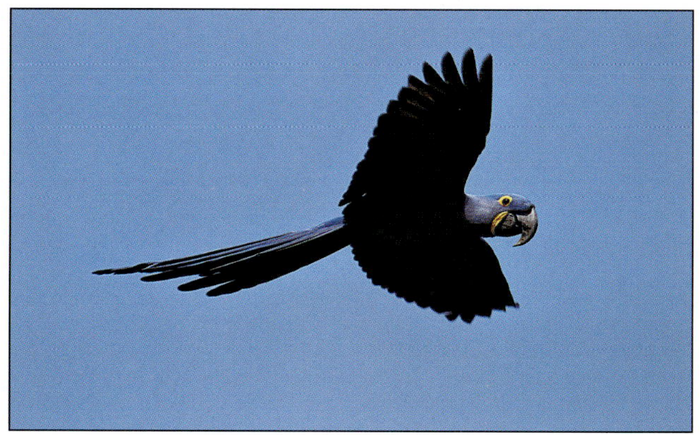

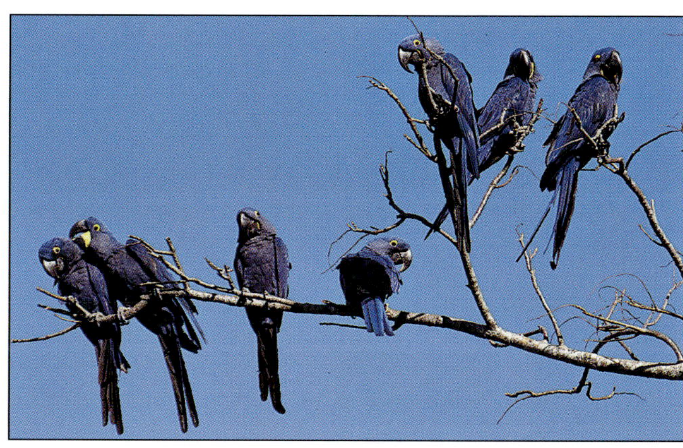

Most often, Hyacinth Macaws are heard long before they are seen as they have to be the nosiest bird in the Pantanal. Viewers have a good chance of seeing this macaw, the largest in the world, because their numbers are increasing. Those chances increase even more if one drives the whole Transpantaneria all the way to Porta Jofre. In fact, if one is near Reserva Ecologica do Jaguar about 100 kilometers down the highway, for a small fee the owner will call a small flock of wild Hyacinth Macaws to within viewing and photographic range. Another area for easy sighting is Porta Jofre where macaws feed on fruit of local palm trees.

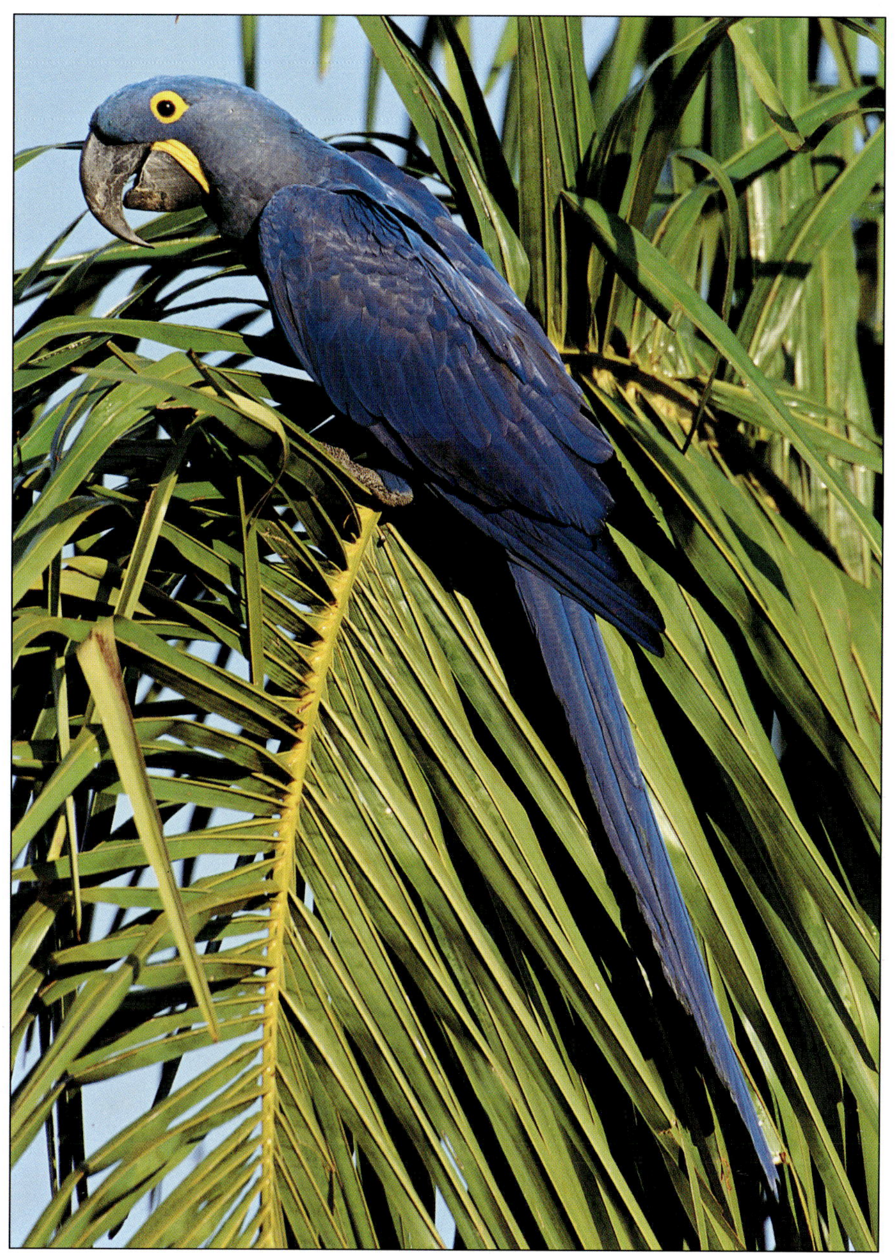

Hyacinth Macaw
Anodorhynchus hyacinthinus

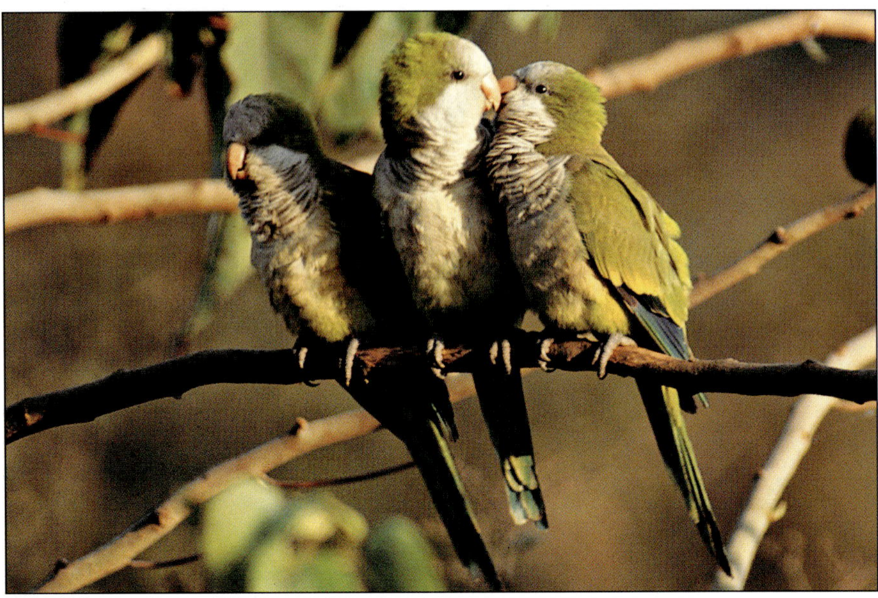

Monk Parakeet
Myiopsitta monachus

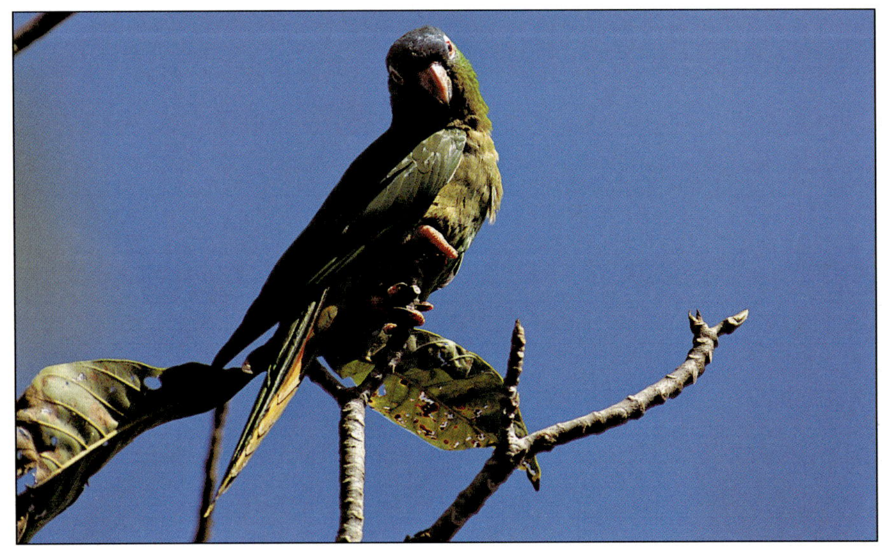

White-eyed Parakeet
Aratinga leucophthalmus

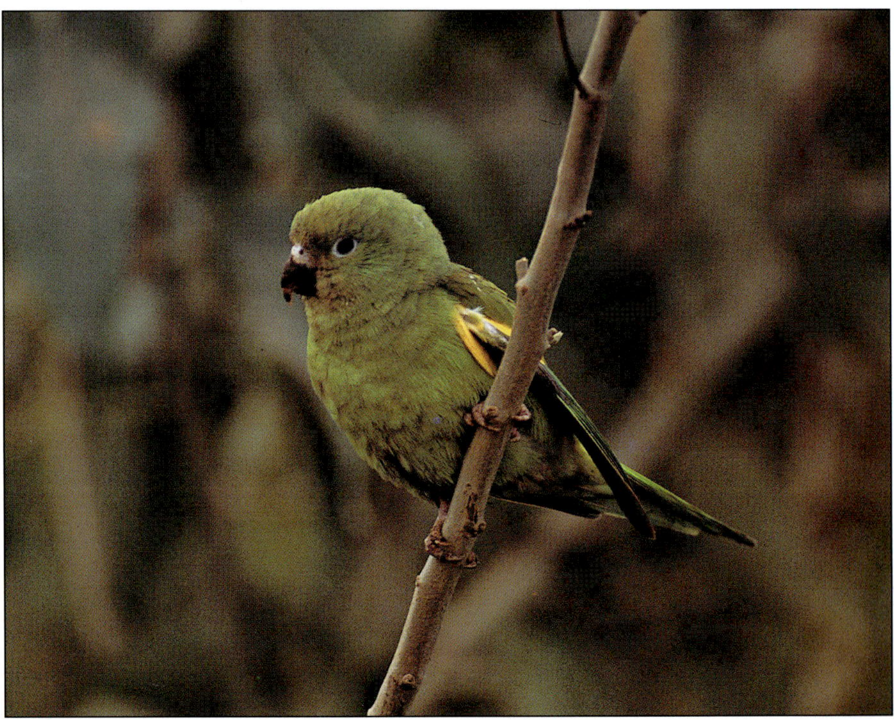

Canary-winged Parakeet
Brotogeris versicolurus

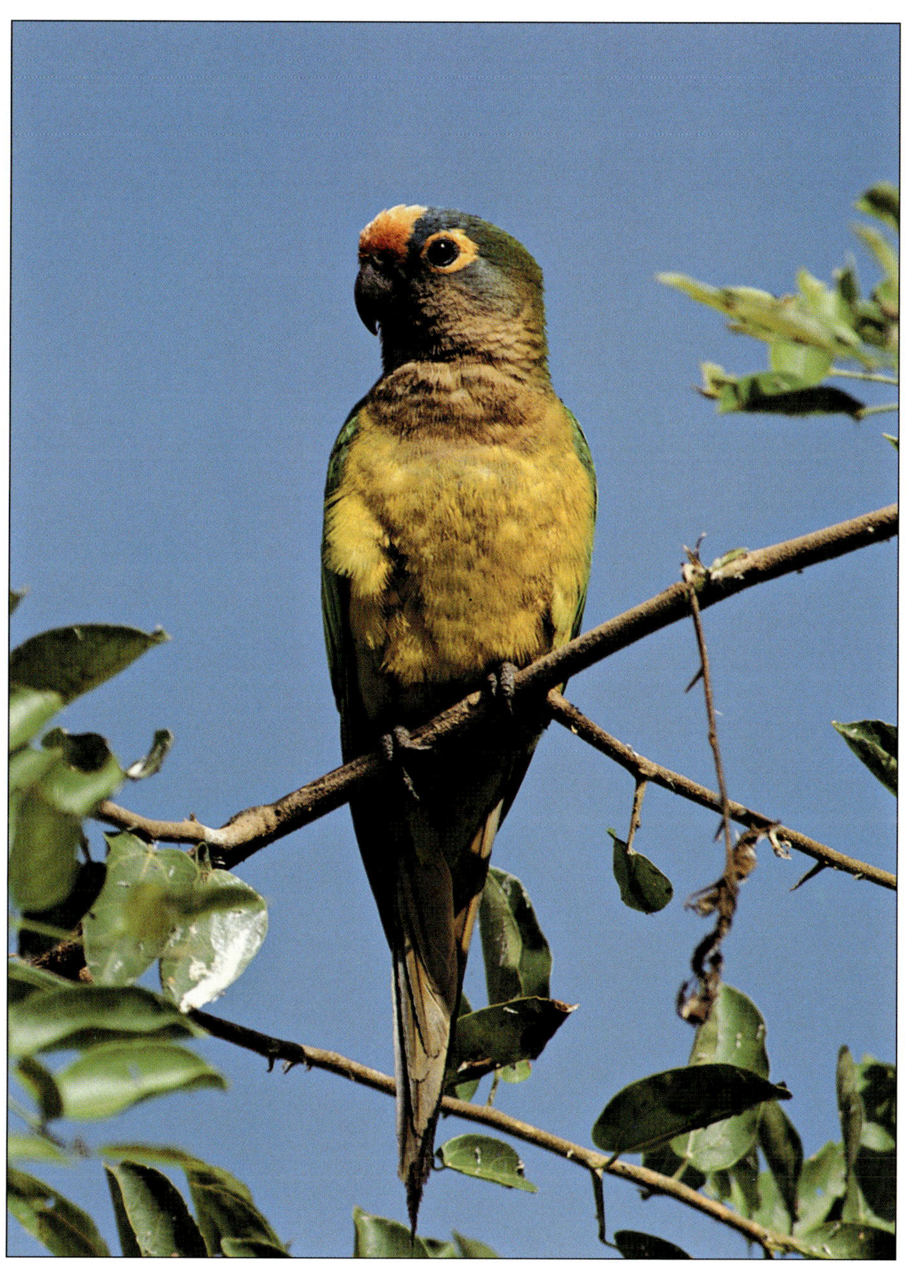

Peach-fronted Parakeet
Aratinga aurea

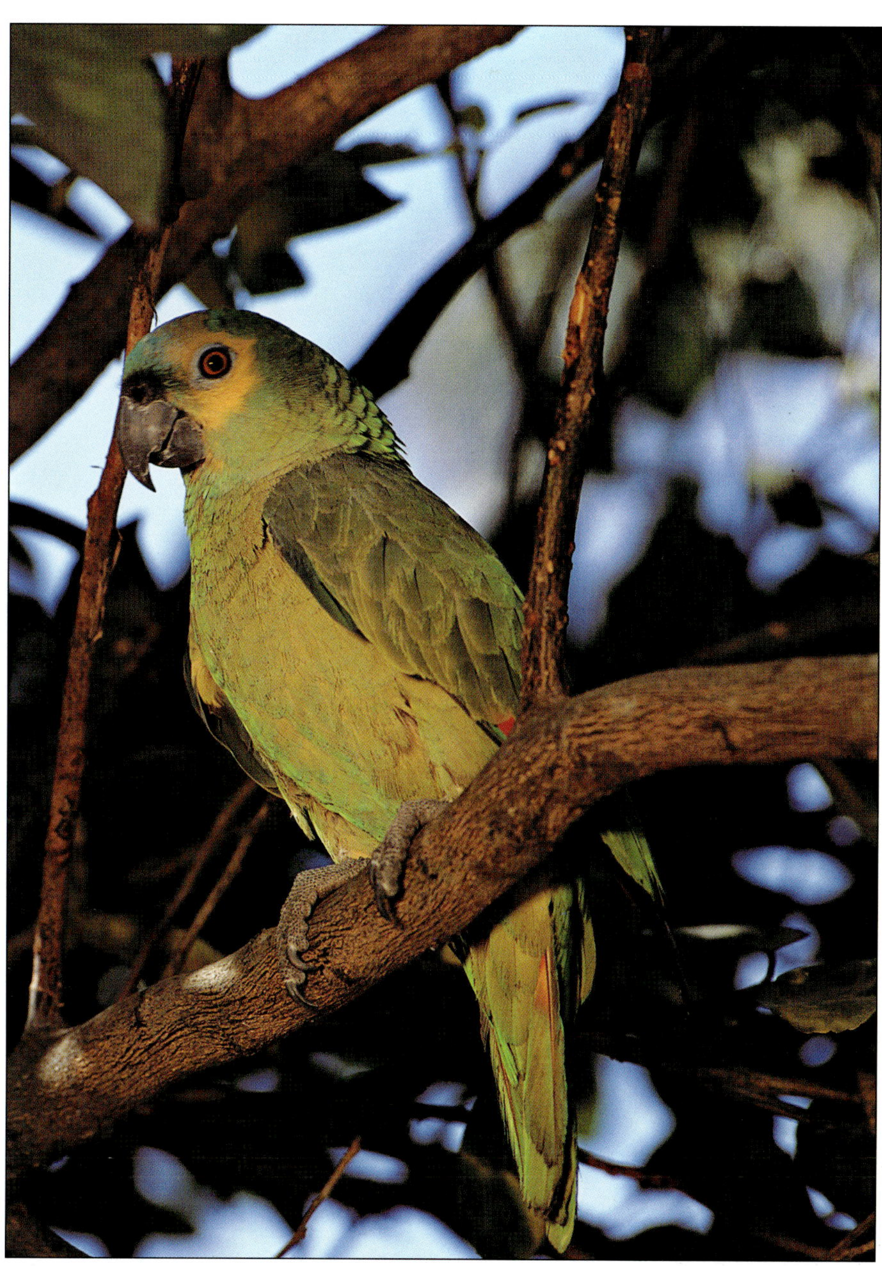

Turquoise-fronted Parrot
Amazona aestiva

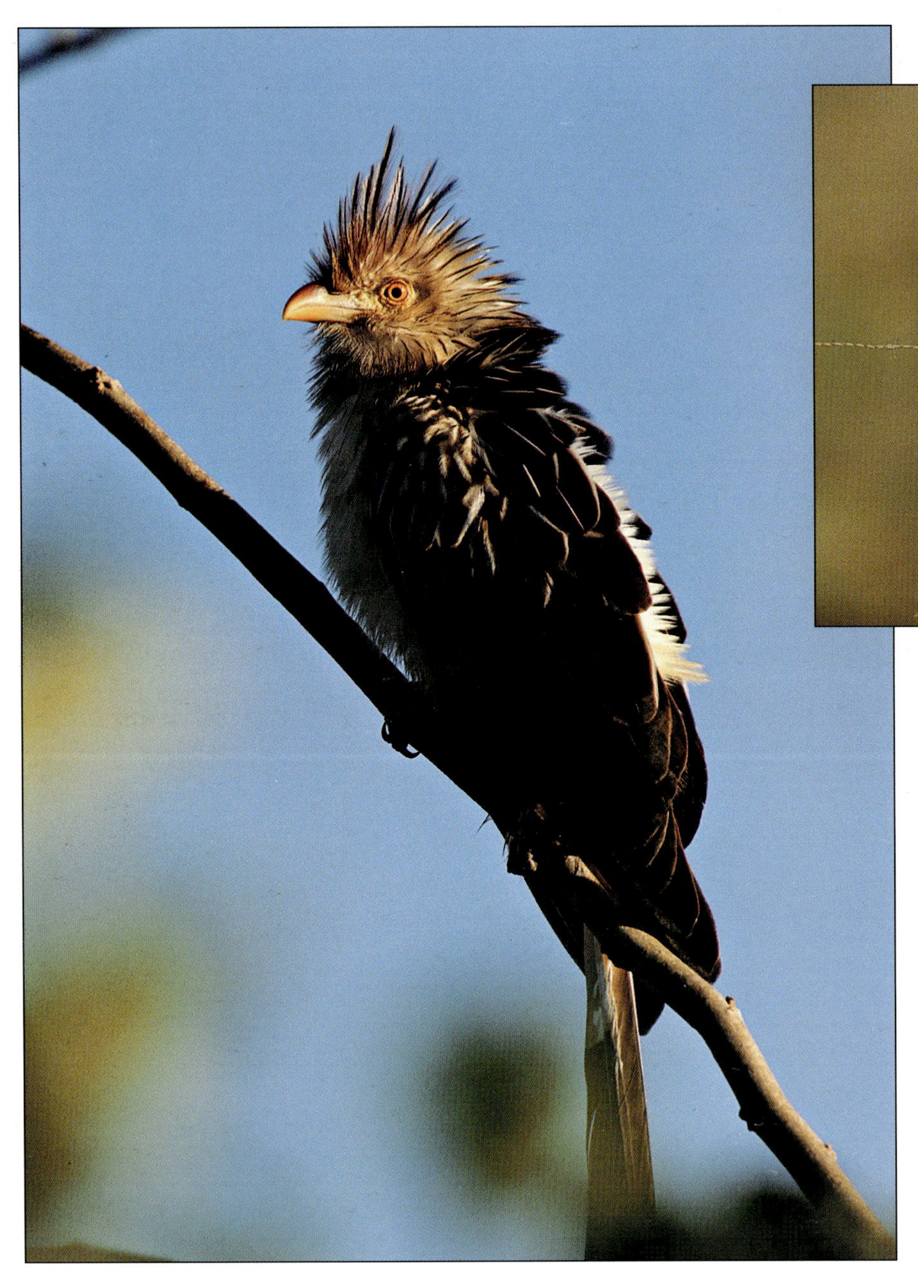

Guira Cuckoo
Guira guira

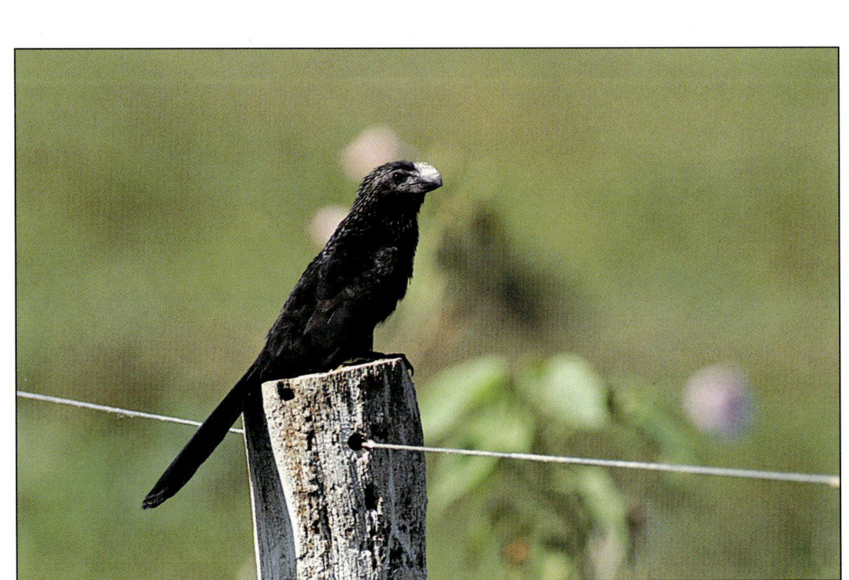

Smooth-billed Ani
Crotophaga ani

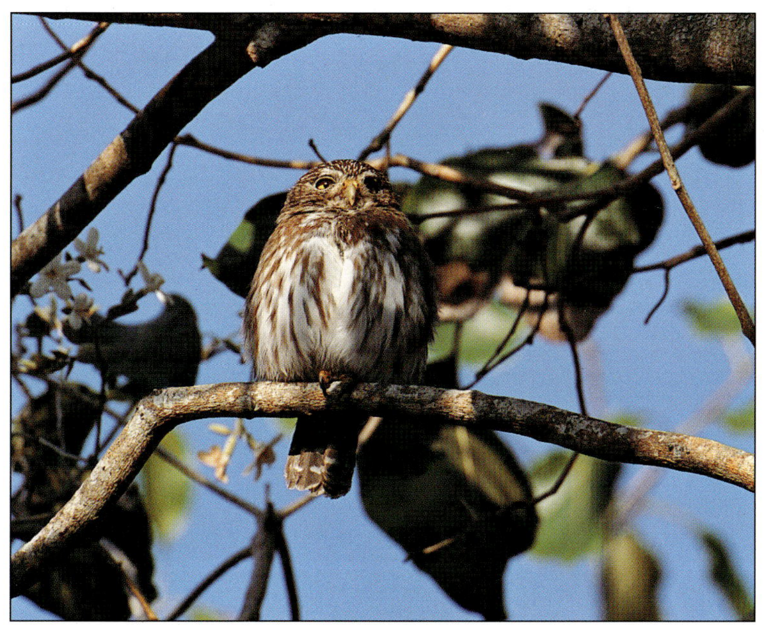

Ferruginous Pygmy Owl (male)

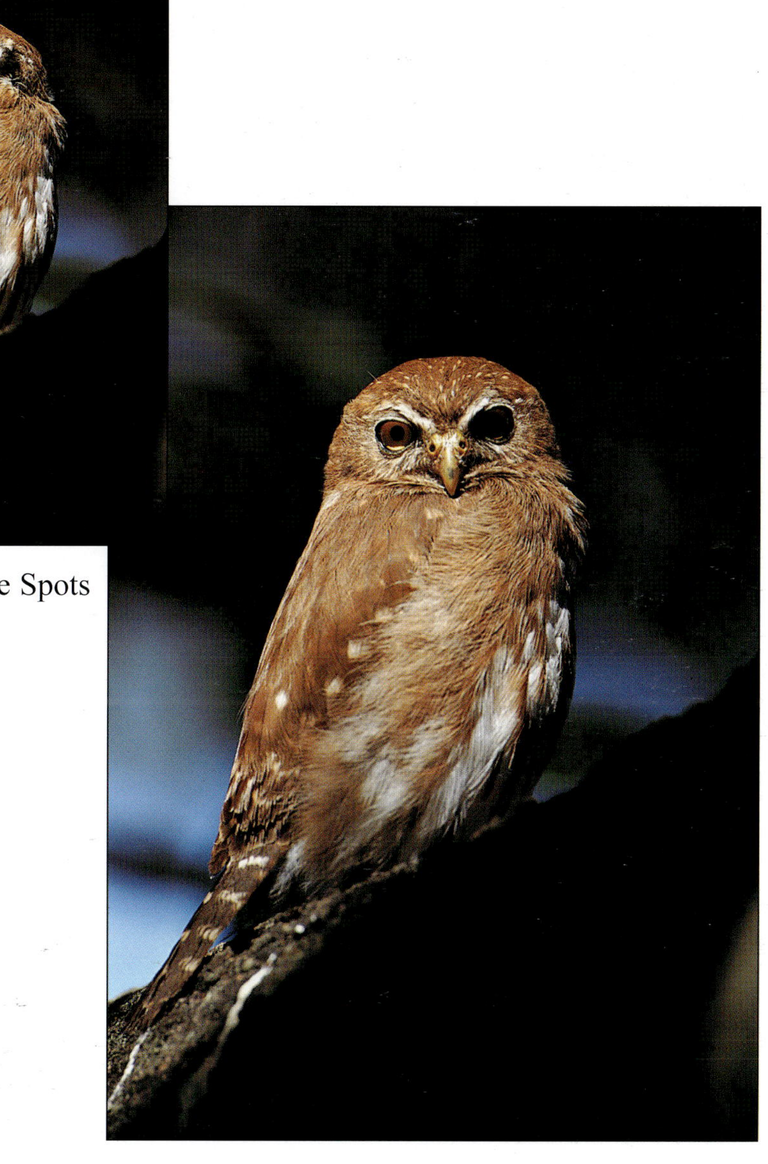

Showing False Eye Spots

The first specimens of Ferruginous Pygmy Owl were collected in Brazil, thus the species name *brasilianum*. The rusty color of the female gives it the common name ferruginous. One familiar characteristic of smaller owls such as this one is a pair of eye spots at the back of the head often thought to deceive predators. Active at night, this little owl can also be seen hunting during the day. While observing one perched, watch for a periodic flick of its tail which is characteristic.

Ferruginous Pygmy Owl (female)
Glaucidium brasilianum

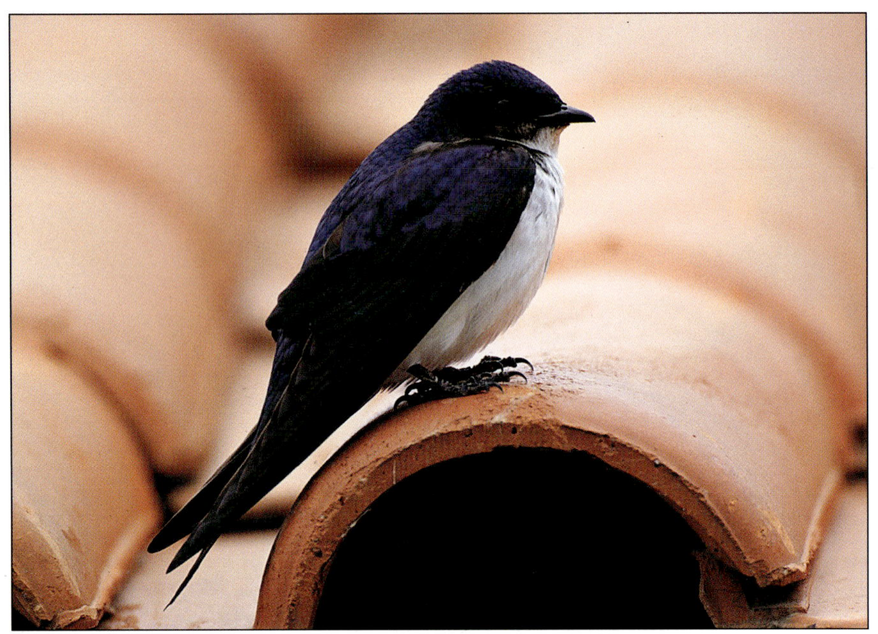

Blue-and-White Swallow
Notiochelidon cyanoleuca

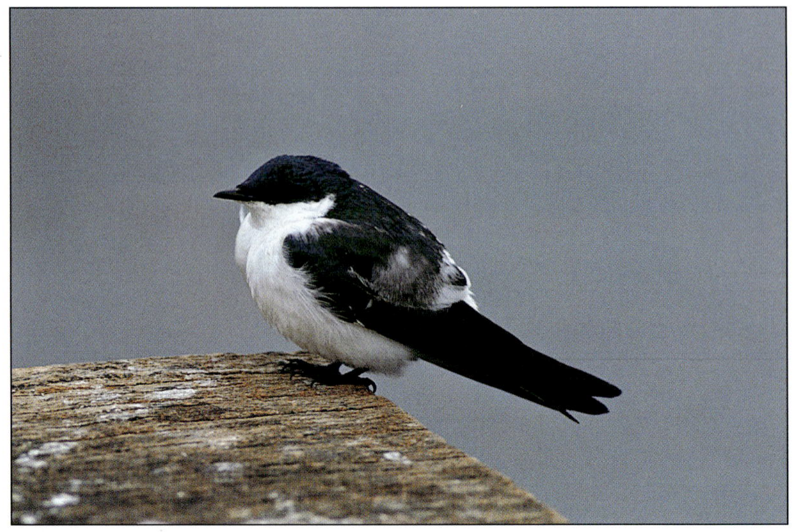

White-winged Swallow
Tachycineta albiventer

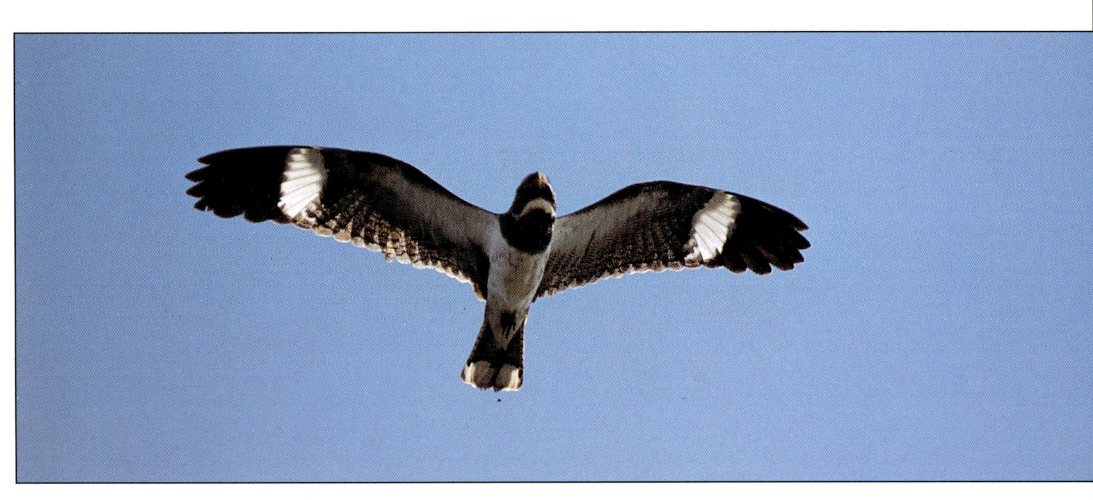

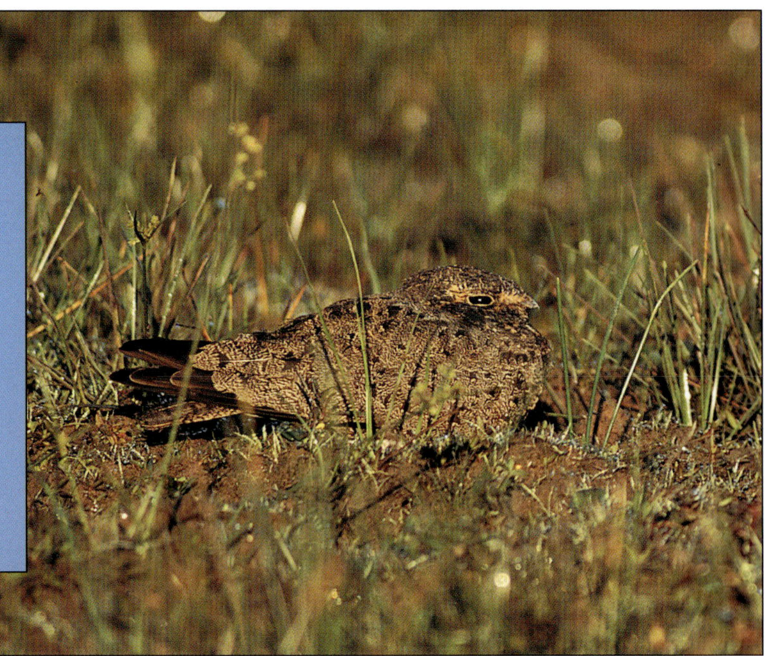

Nacunda Nighthawk
Podager nacunda

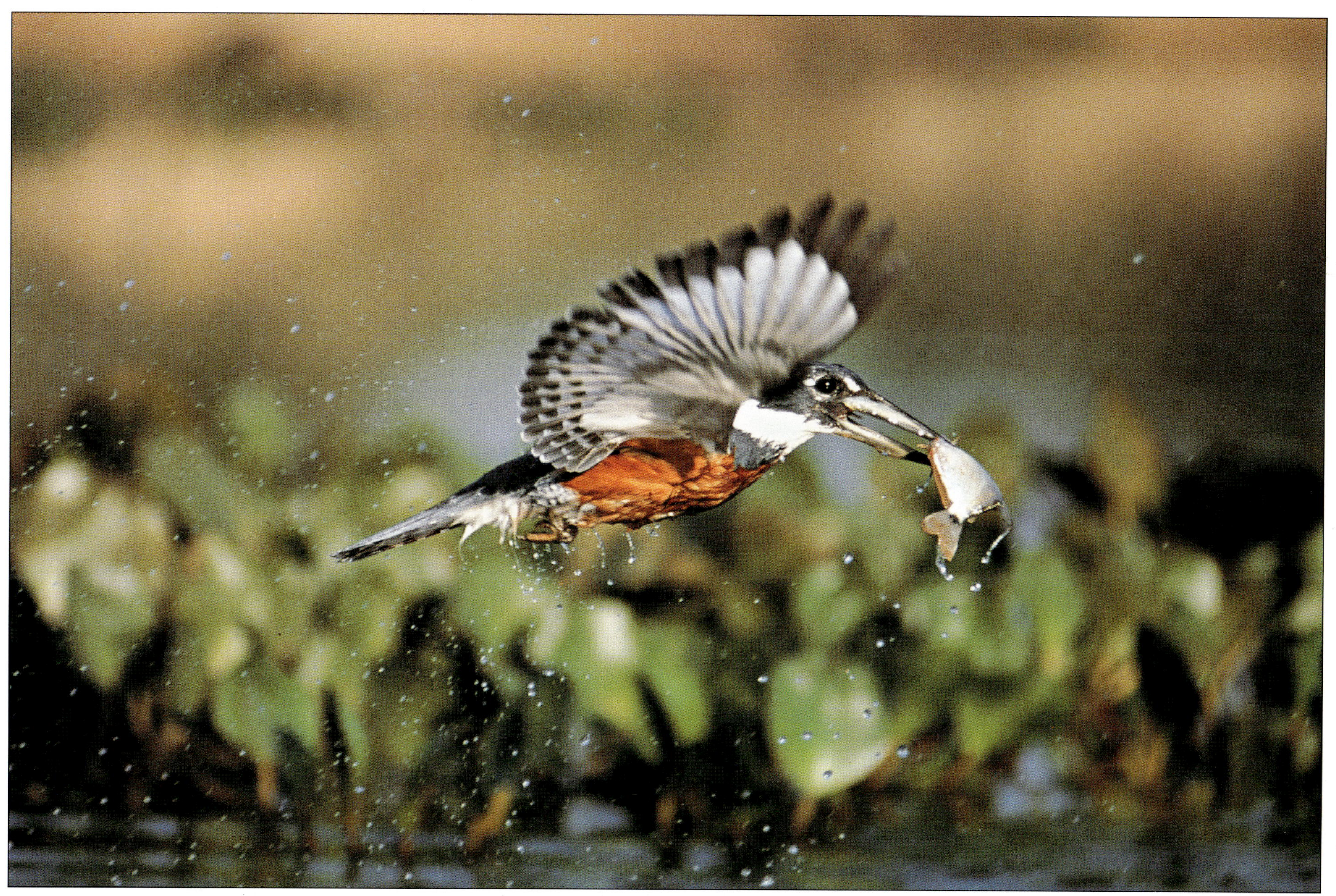

Ringed Kingfisher
Ceryle torquata

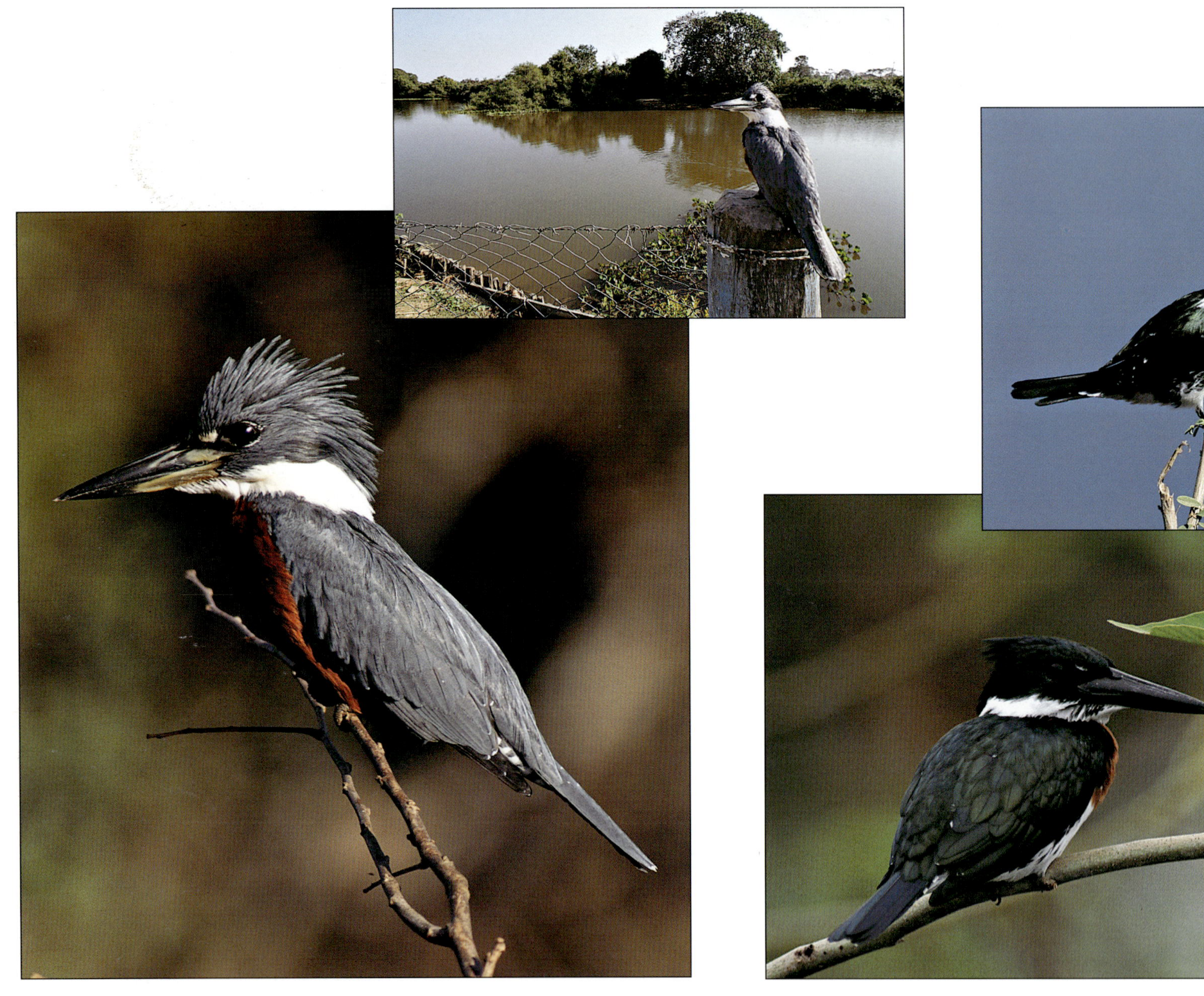

Ringed Kingfisher
Ceryle torquata

Amazon Kingfisher
Chloroceryle amazona

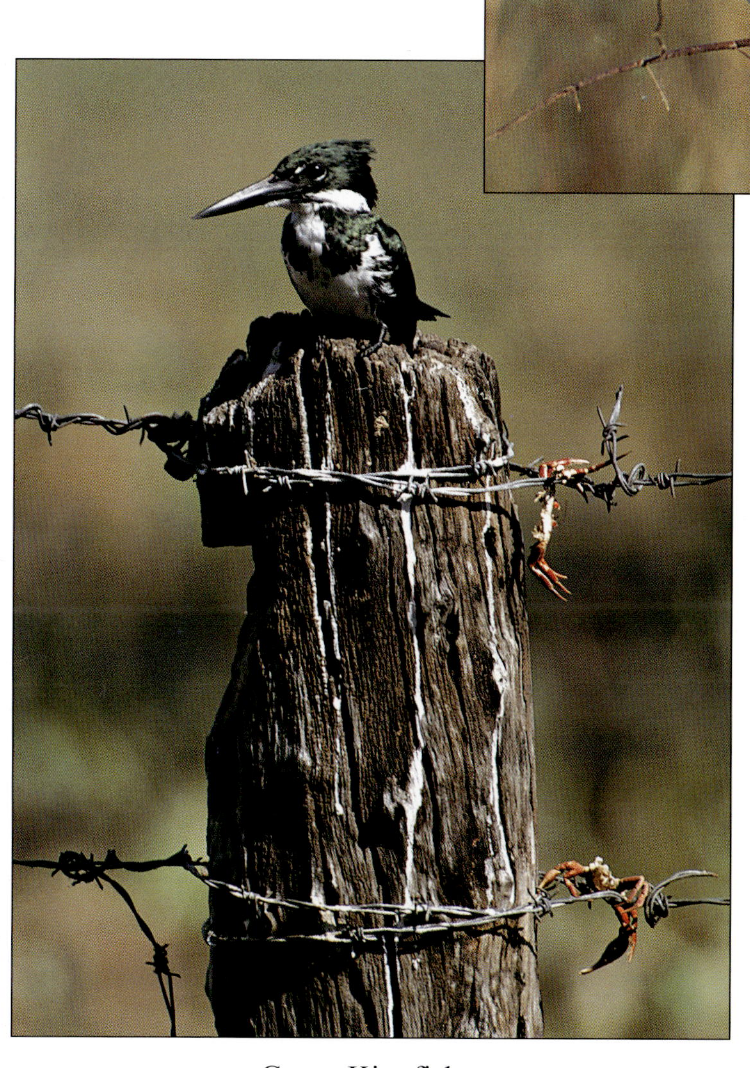

Green Kingfisher
Chloroceryle americana

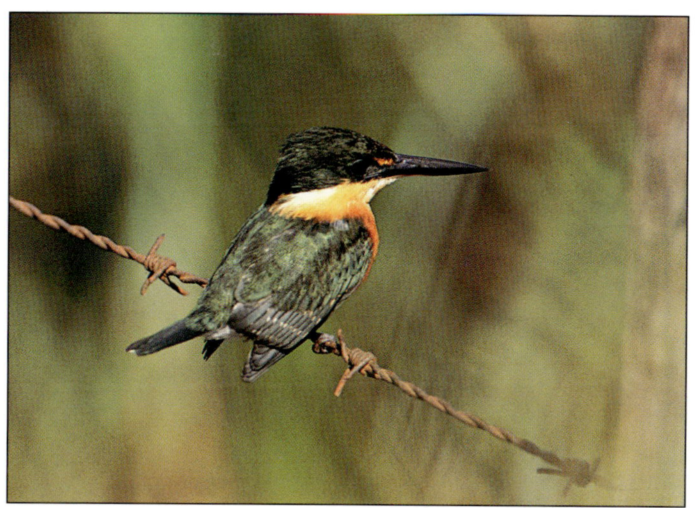

Pygmy Kingfisher
Chloroceryle aenea

One can readily observe three species of kingfishers during a visit to the Pantanal. The Ringed Kingfisher is the most common followed in lower numbers by the Amazon Kingfisher. Present in even fewer numbers is the Green Kingfisher. If one is lucky, occasionally it is possible to spot a fourth species, the Pygmy Kingfisher. Trying to photograph the pygmy is difficult at best. Due to this bird's small size, a long telephoto lens is needed. Even with the long lens, one must be as close as possible to get a good photo.

Any bridge along the Transpantaneria will host minimally a few pairs of kingfishers. They perch on the bridge or on favorite branches, carefully watching the water below. As quick as lightning, they drop with cunning accuracy, breaking the water's surface and snatching an unsuspecting minnow or small fish. Then they fly back to that favorite perch and swallow the victim whole and head first.

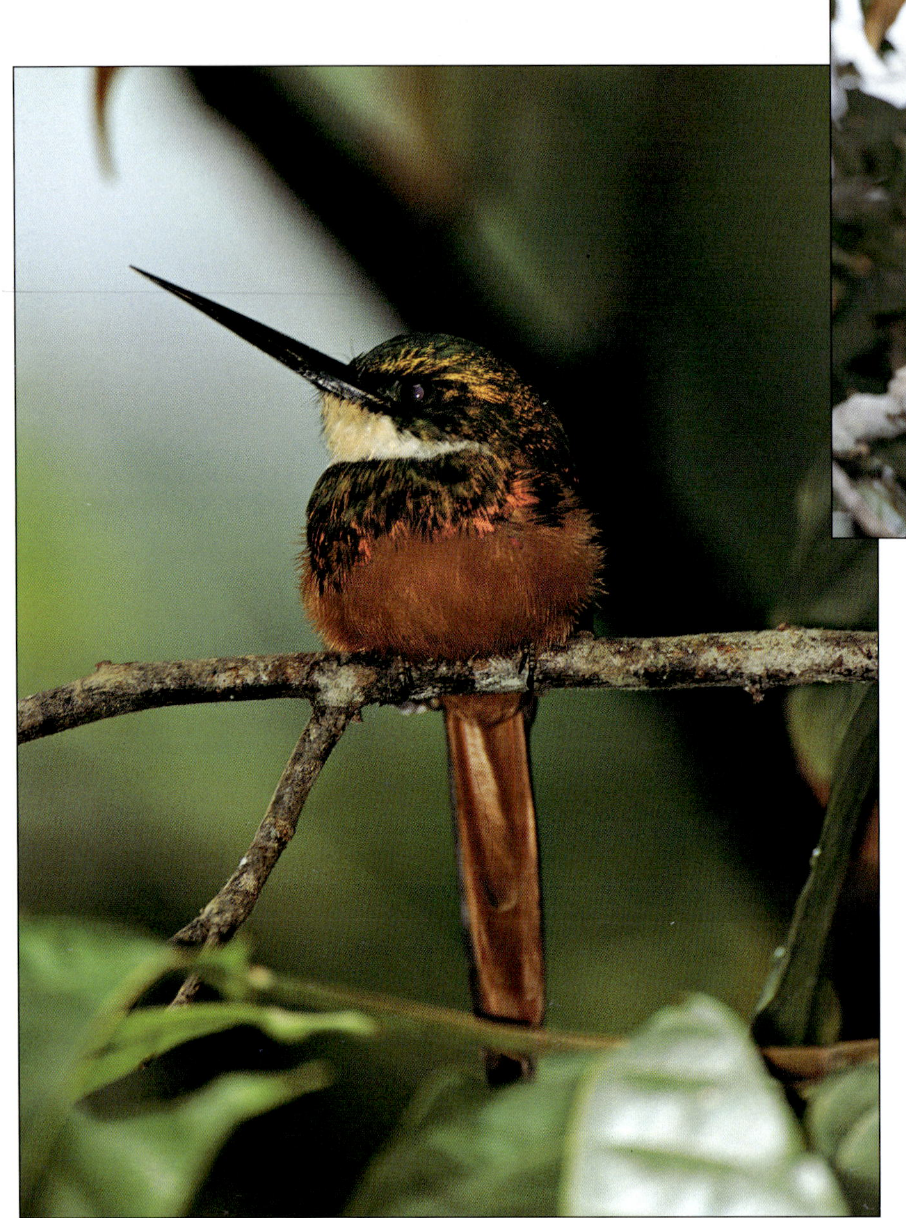

Rufous-tailed Jacamar
Galbula ruficauda

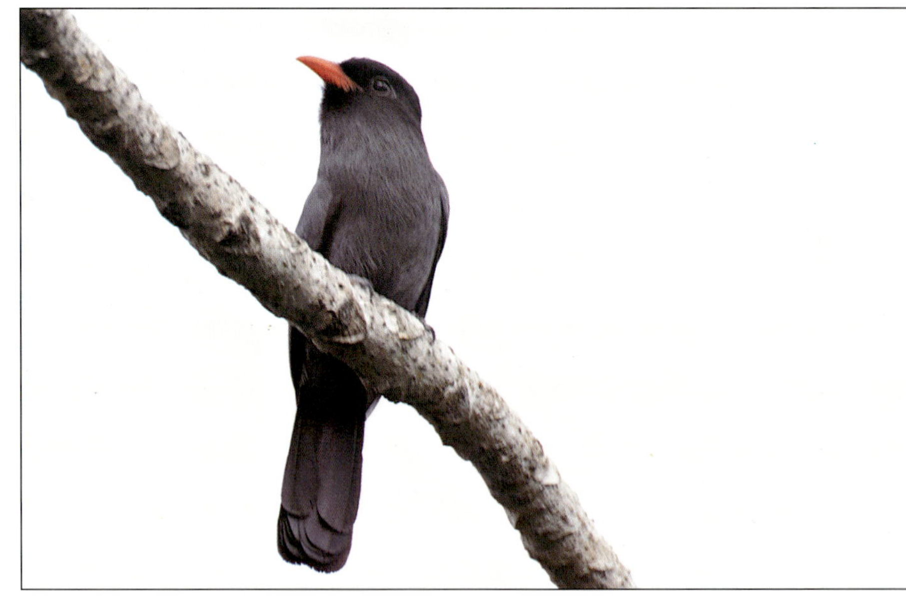

Black-fronted Nunbird
Monasa nigrifrons

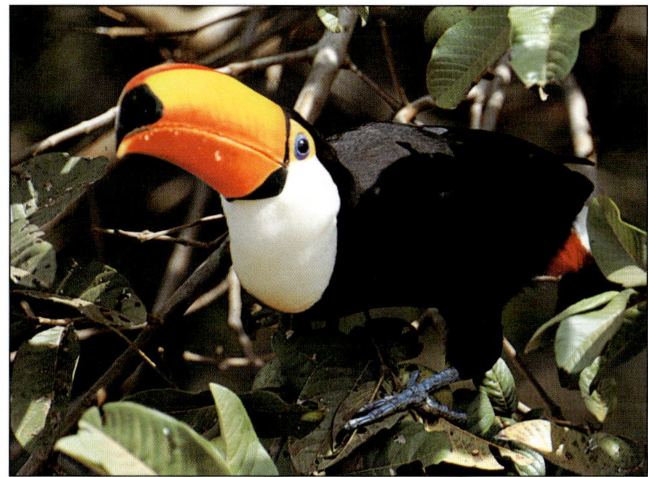

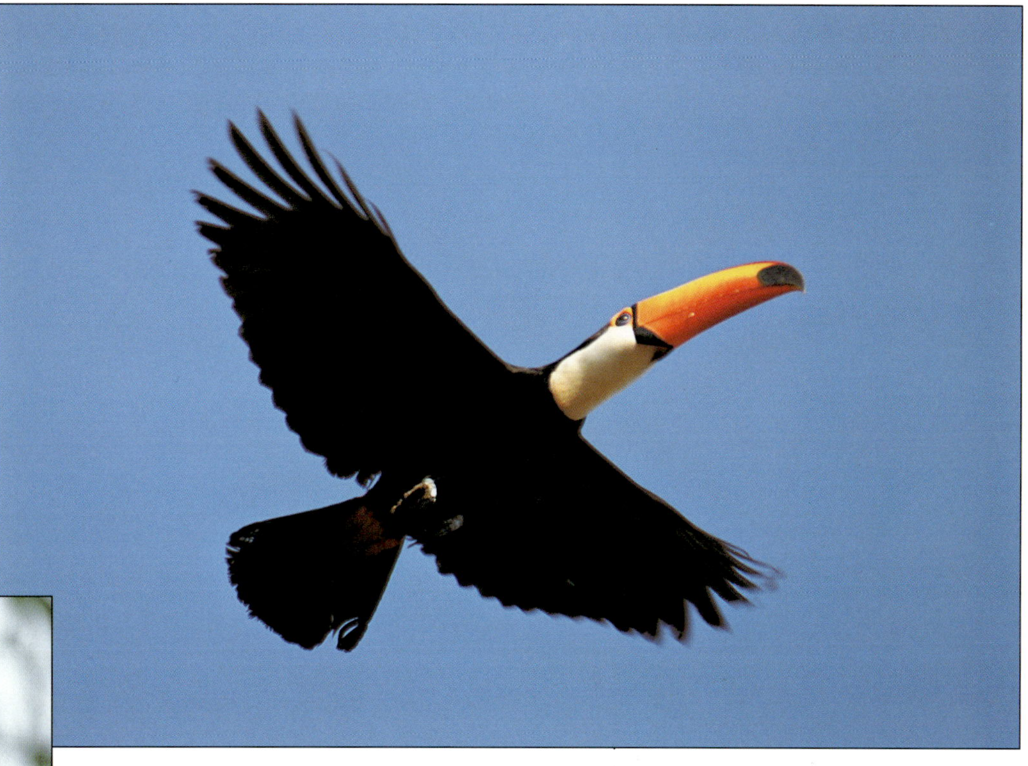

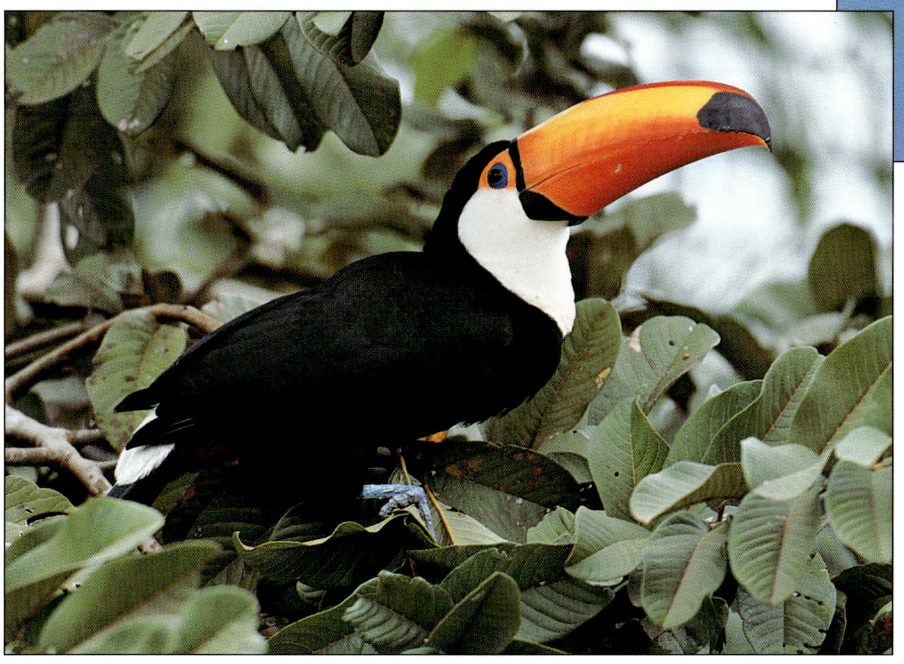

Toco Toucan
Ramphastos toco

One of the more comical looking birds in the world is the toucan. Only found in Central and South America, they are distantly related to hornbills of Africa. The largest, most common one found in the Pantanal is the Toco Toucan. It is more readily seen as it will often leave the protection of forest and fly across open fields and marshy areas. The large colorful bill, the purpose of which to tear through leathery skins of many fruits, easily identifies them.

Another group, often confused with toucans because they also have colorful bills, is the aracaries. They are classified separately as their bills tend to be somewhat smaller in size.

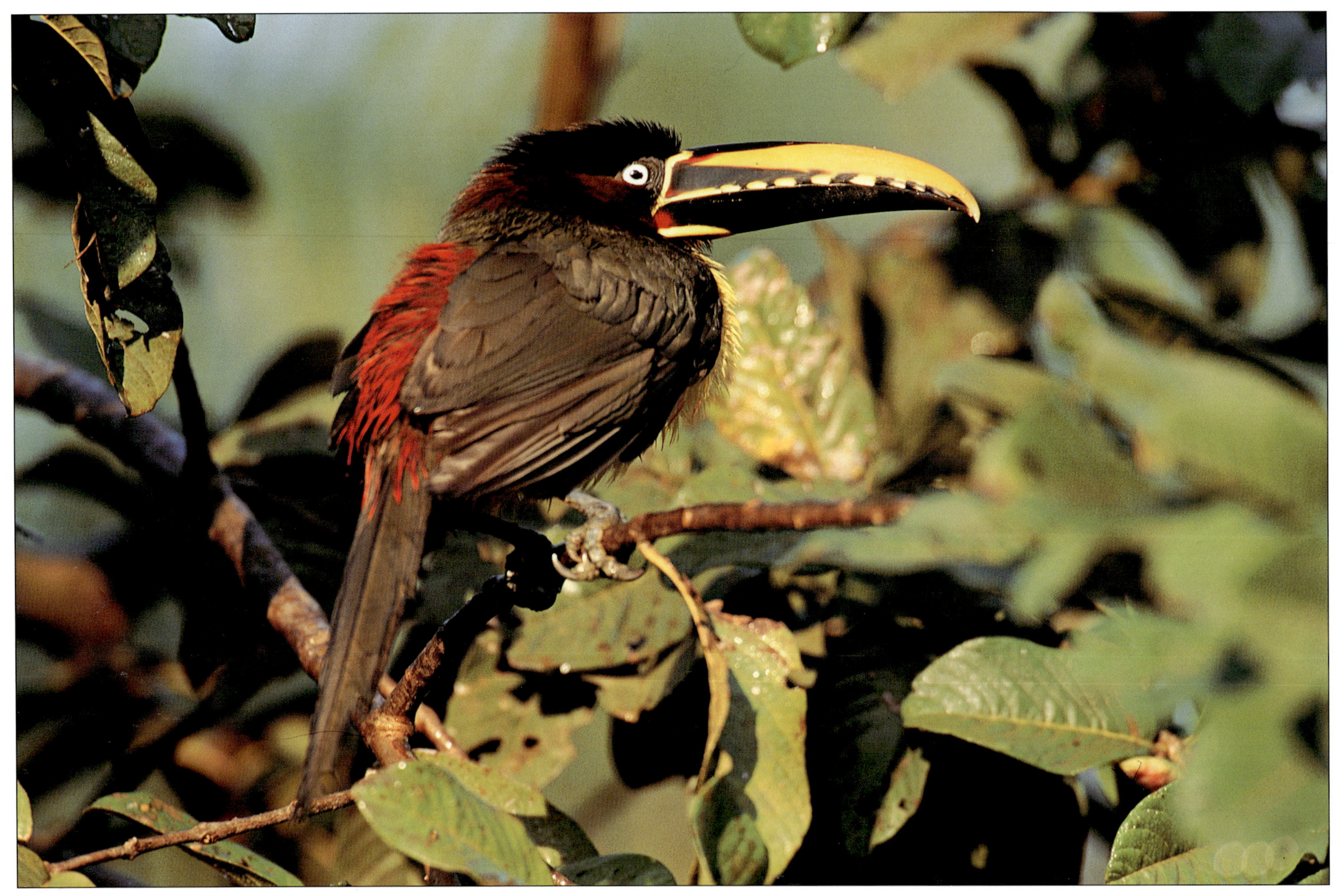

Chestnut-eared Aracari
Pteroglossus castanotis

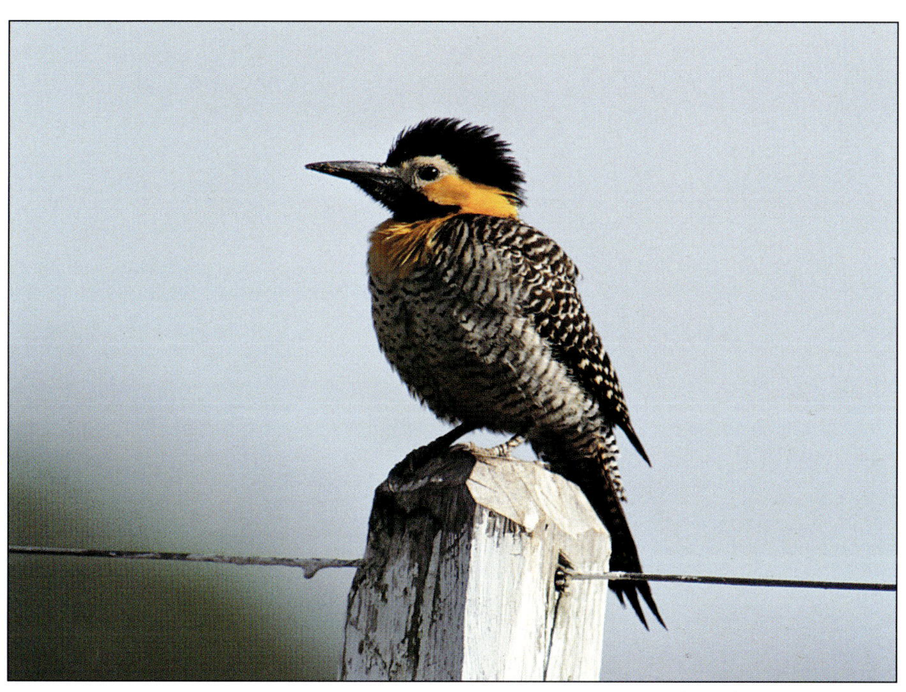

Campo Flicker
Colaptes campestris

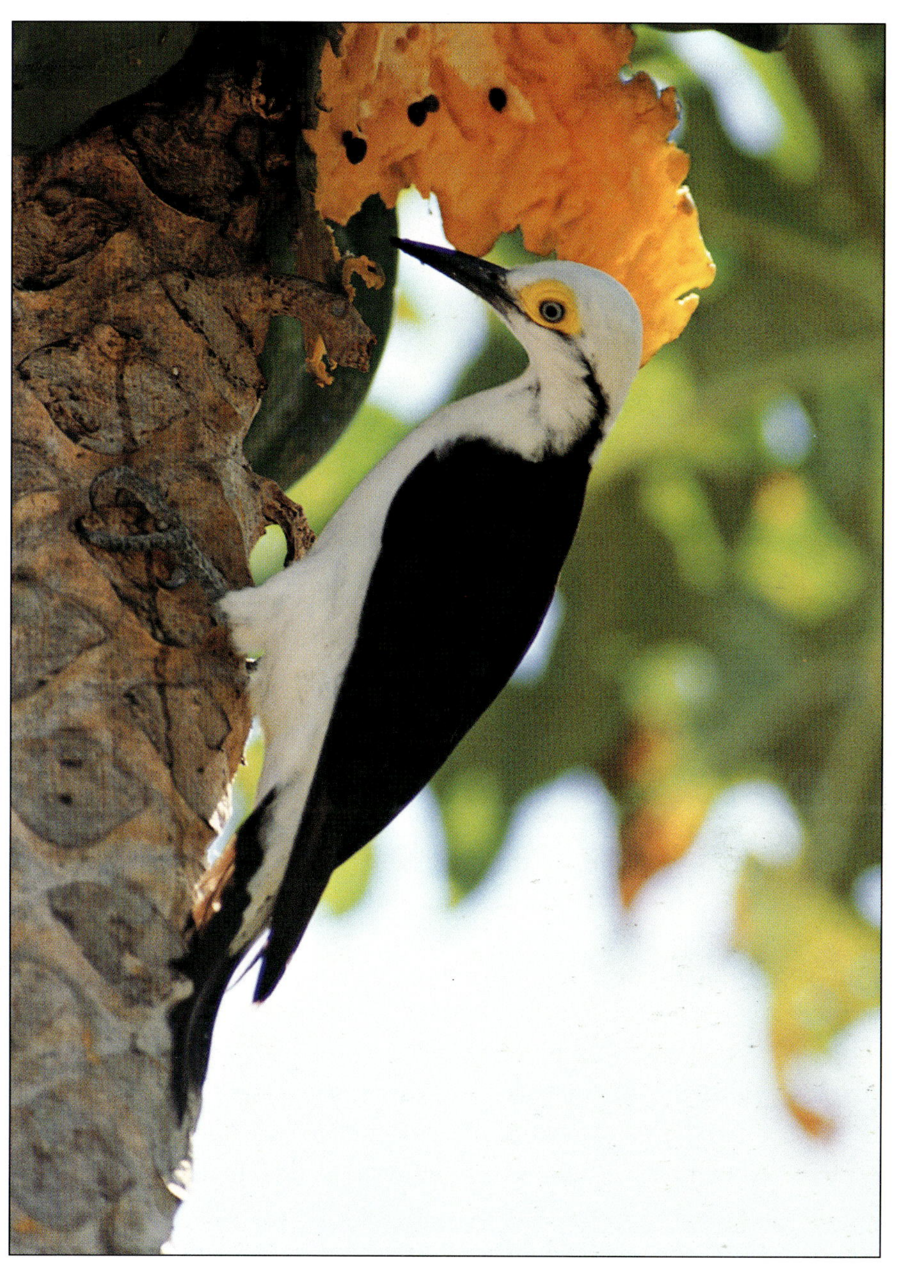

White Woodpecker
Leuconerpes candidus

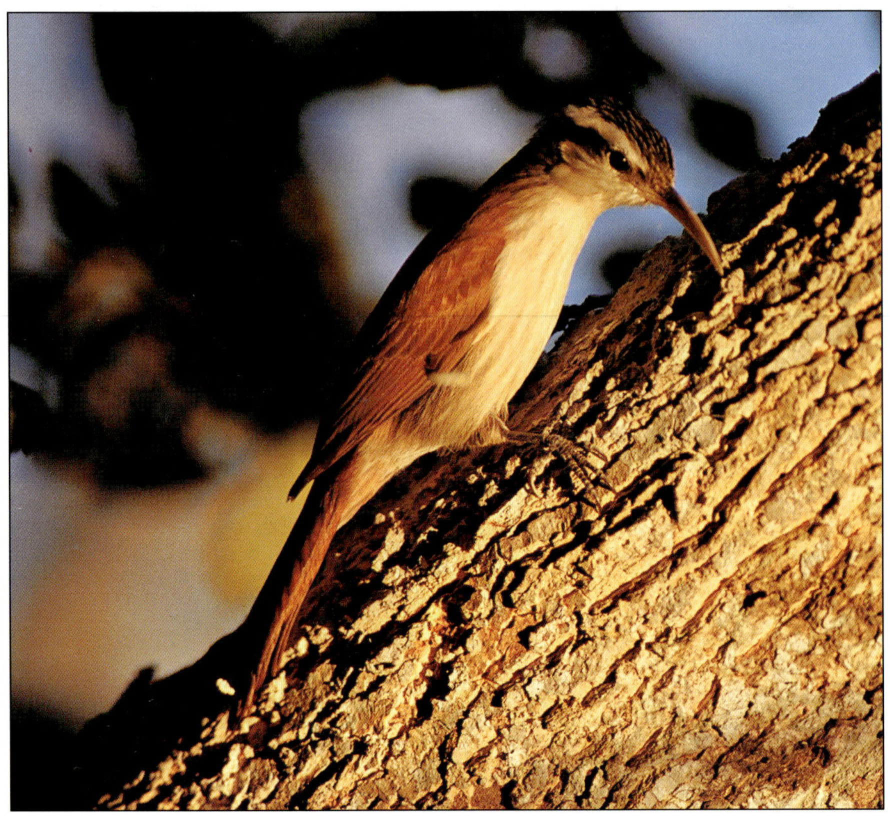

Narrow-billed Woodcreeper
Lepidocolaptes angustirostris

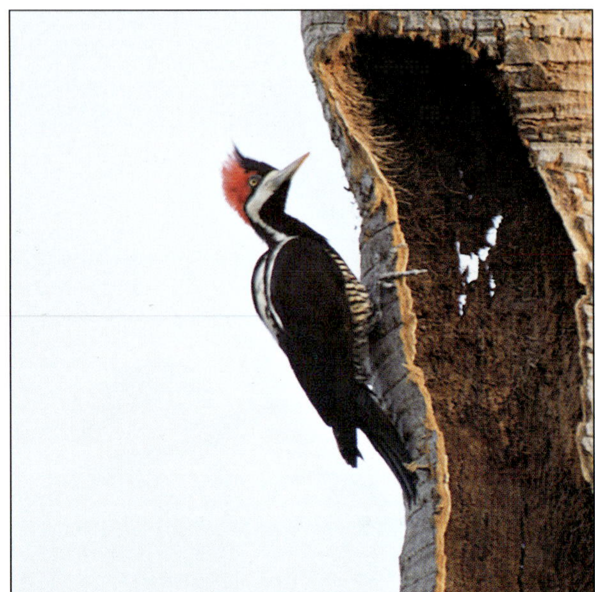

Lineated Woodpecker
Dryocopus lineatus

Woodpeckers and creepers have adapted to life on tree trunks and branches more successfully than most other birds. It is easy for them to hold a heads-up perched position on any vertical tree trunk surface. Here they investigate for insects or larva of insects by pecking the bark away or into the wood itself. Their ability to peck without causing any injuries or damage is due to a thick-walled skull and a narrow space between the tough outer membrane of the brain and the brain itself, both of which absorb any shock.

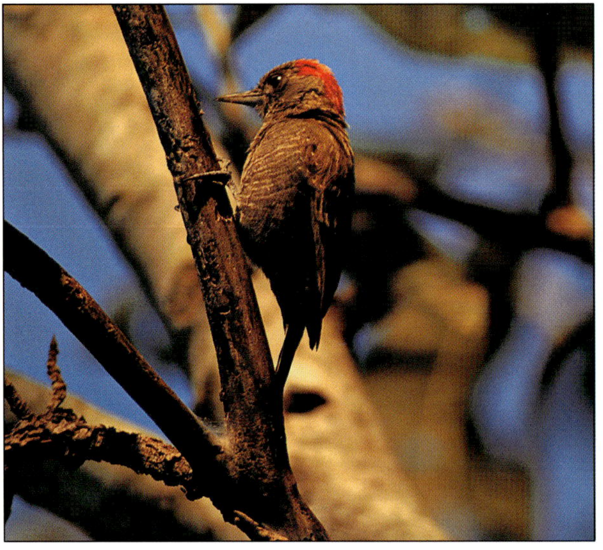

Little Woodpecker
Veniliornis passerinus

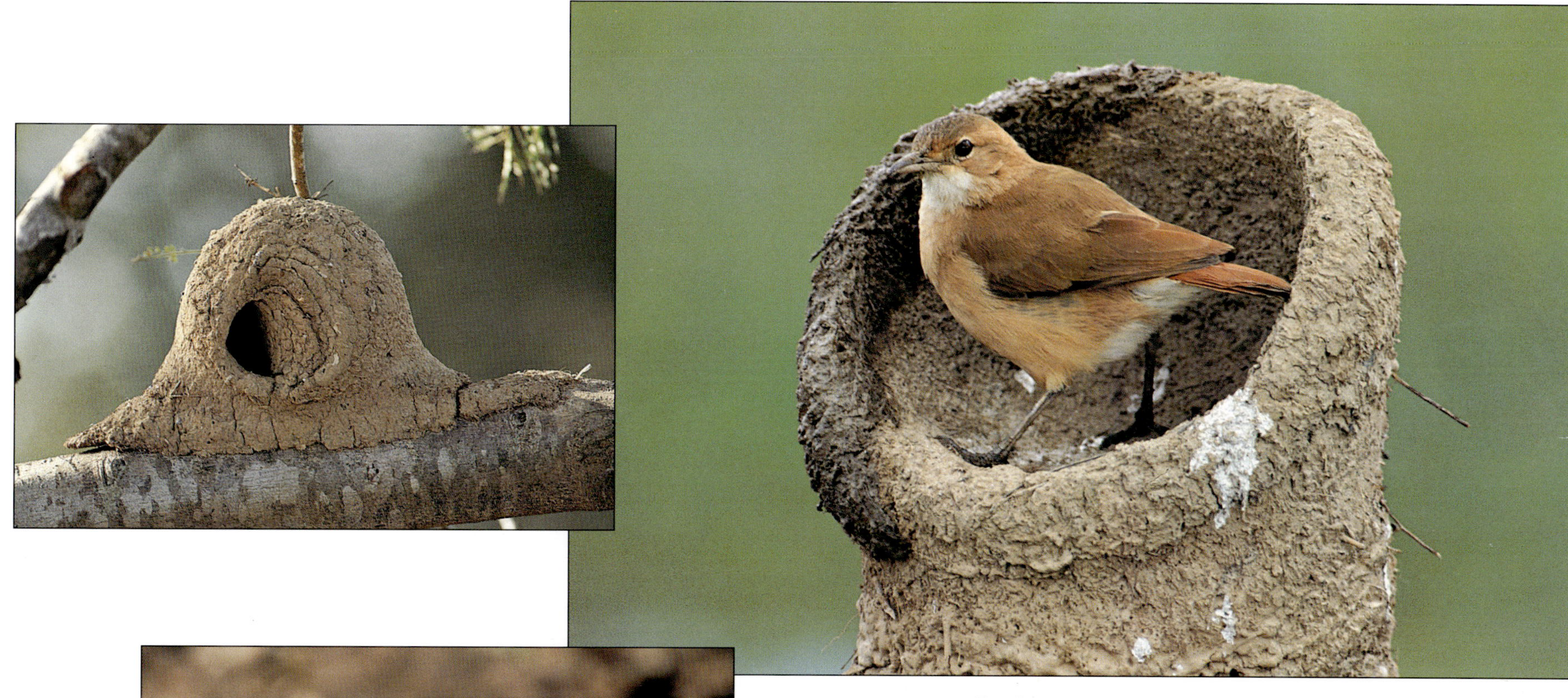

Rufous Hornero
Furnarius rufus

There is one bird that tends to leave its mark on the landscape more than any other. The elaborate adobe mud nests of the Rufous Hornero are only used once, but because of their durability they tend to stick around for years. It takes months to slowly work the wet mud and cow dung into these fortified nests. Sometimes, locations come into short supply and it is not uncommon for horneos to build atop older nests forming condos two or three nests high. Often, these birds are called "oven birds" because of the nest appearing similar to clay dutch ovens, but Rufous Hornero is a more proper name.

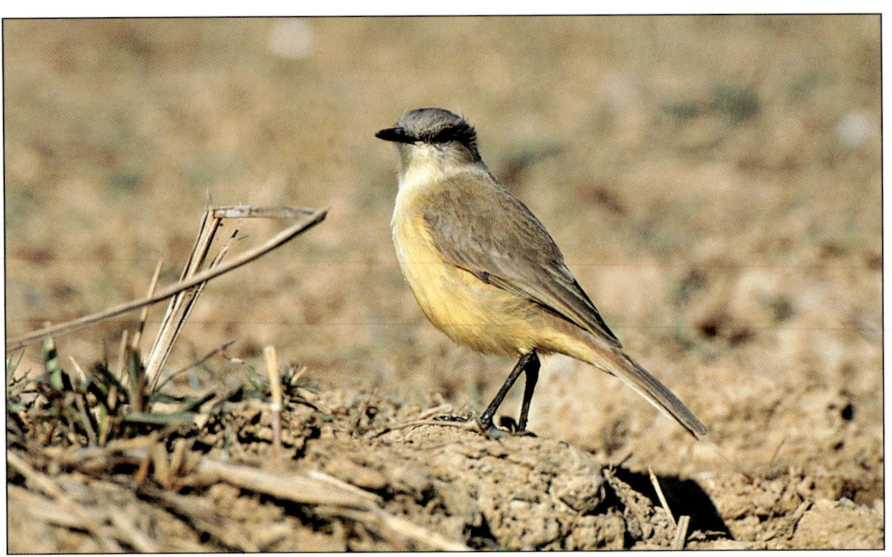

Cattle Tyrant
Machetornis rixosus

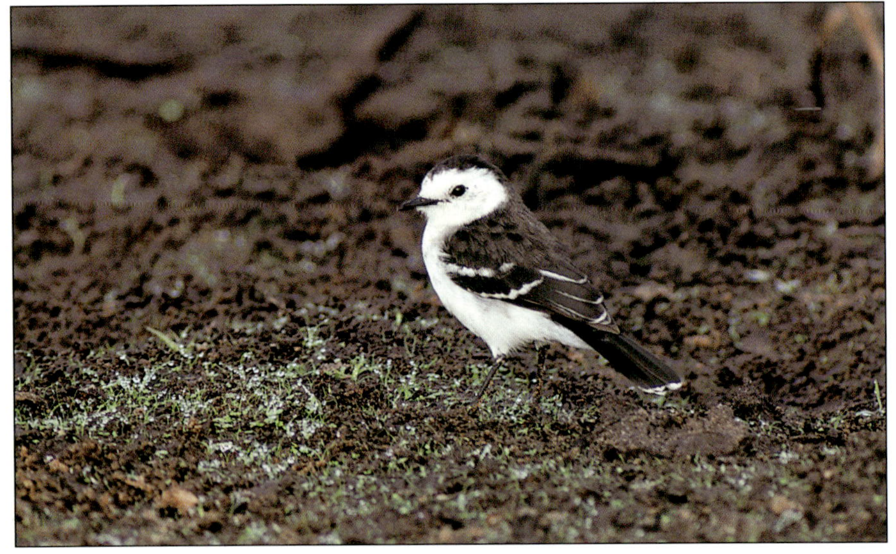

White-bared Water-Tyrant
Fluvicola albiventris

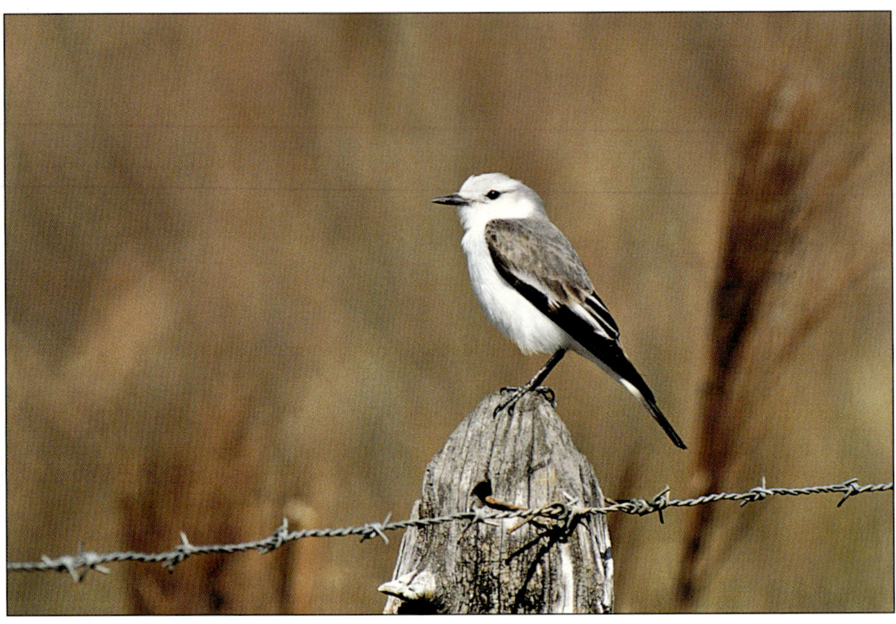

White-rumped Monjita
Xolmis velata

Two birds, which are similar in appearance, are often confused. The color patterns of the Tropical Kingbird and the Cattle Tyrant are almost a perfect match. However, another physical characteristic and a behavior pattern help to separate them. The Cattle Tyrant has long legs and scurries around on the ground picking off insects. The kingbird has shorter legs and loves to perch as it watches patiently for flying insects to snatch in midair.

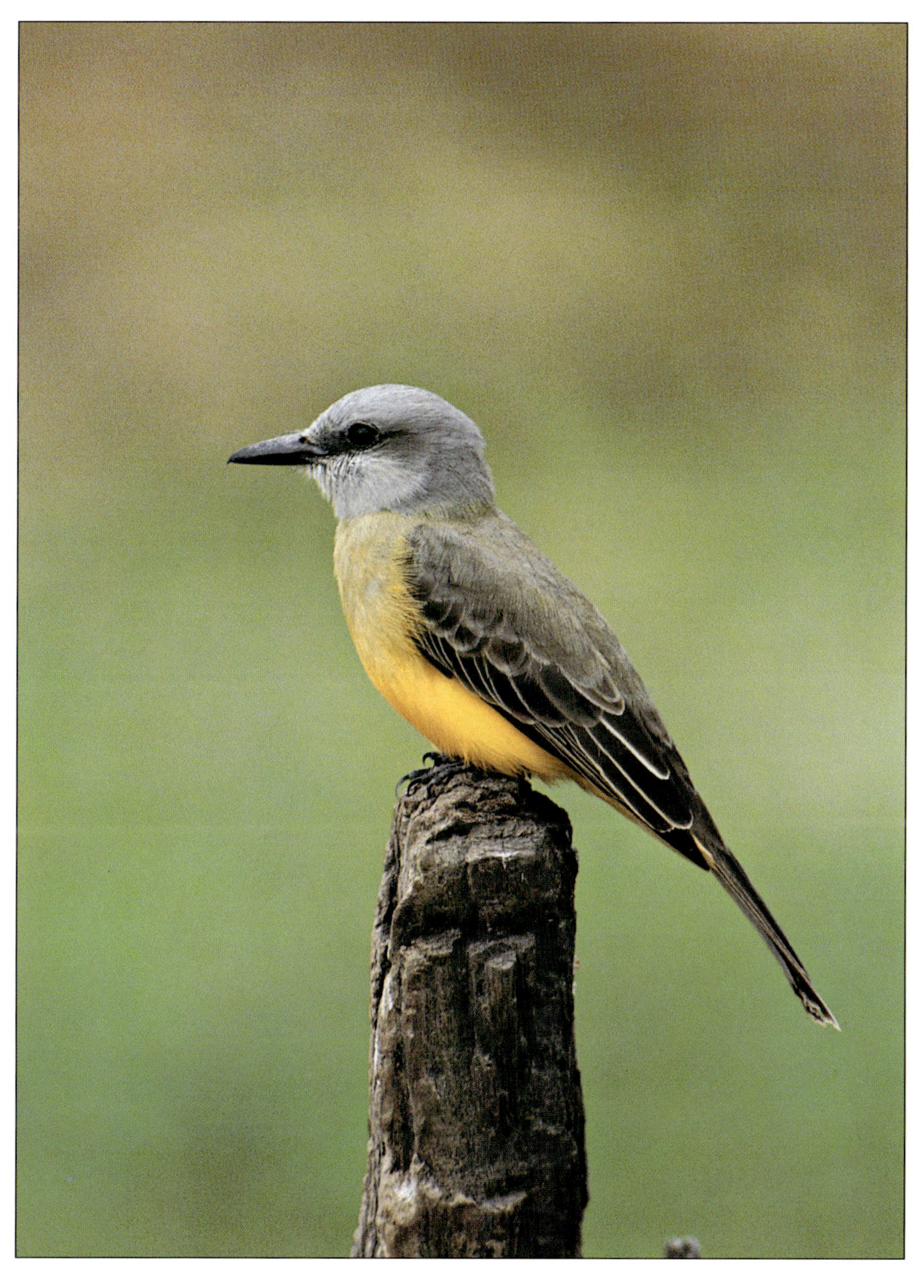

Tropical Kingbird
Tyrannus melancholicus

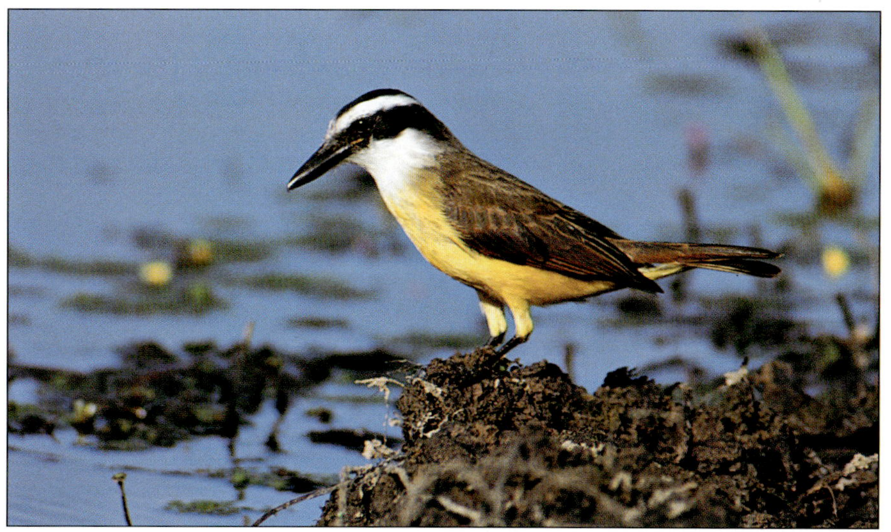

Great Kiskadee
Pitangus sulphuratus

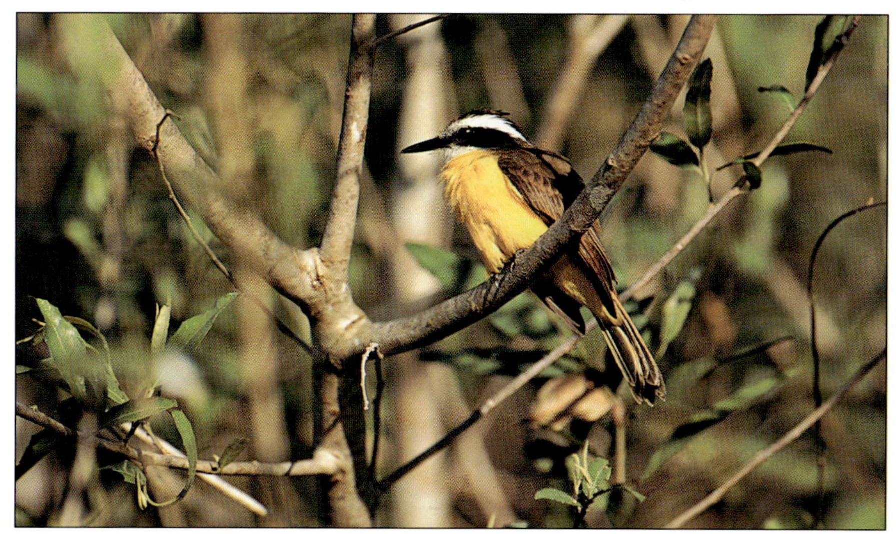

Lesser Kiskadee
Pitangus lictor

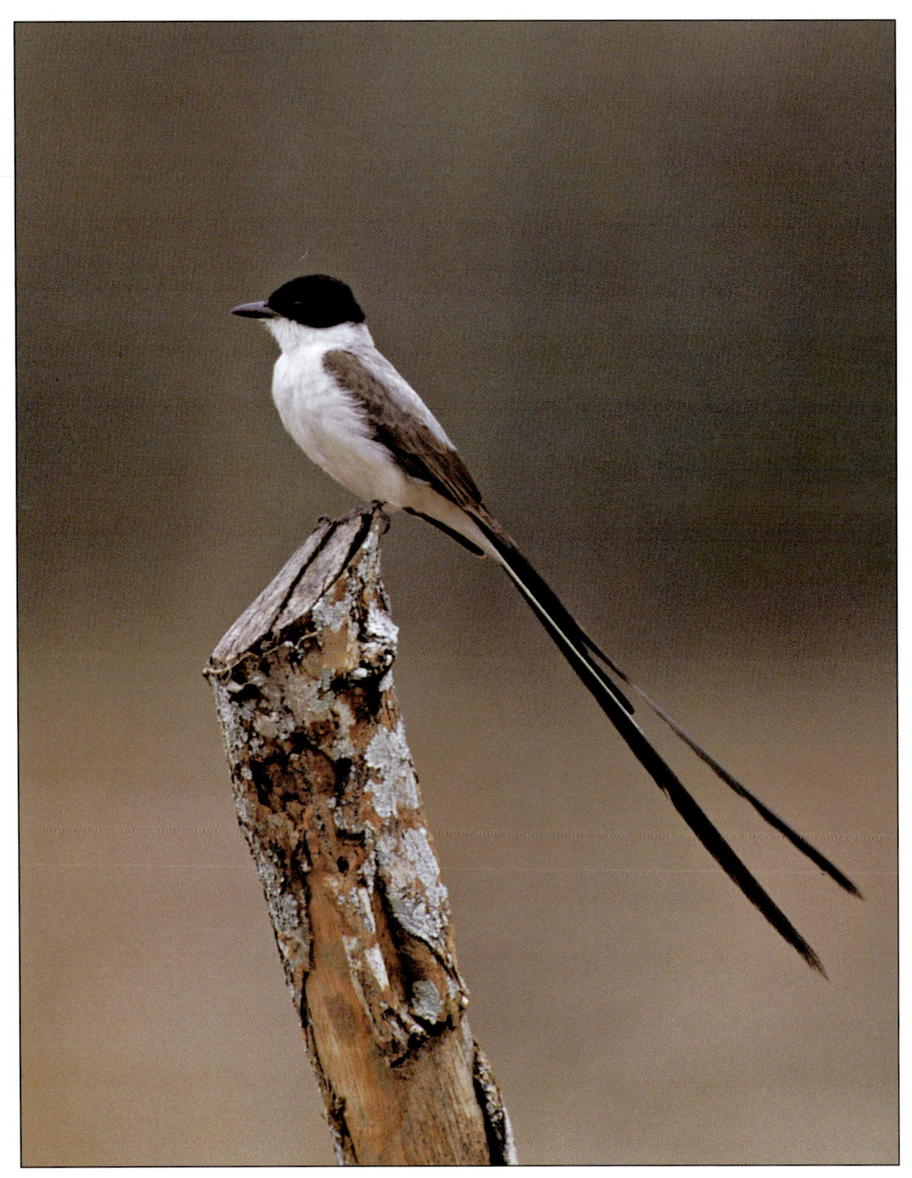

Fork-tailed Flycatcher
Muscivora tyrannus

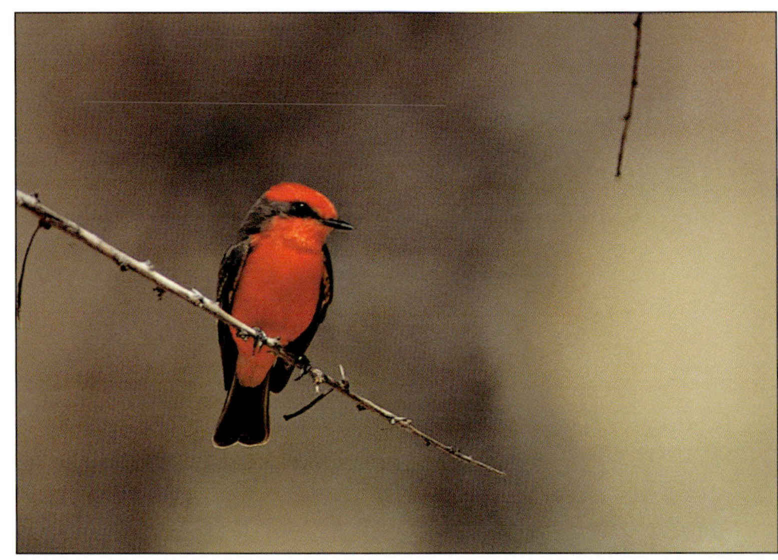

Vermilion Flycatcher
Pyrocephalus rubinus

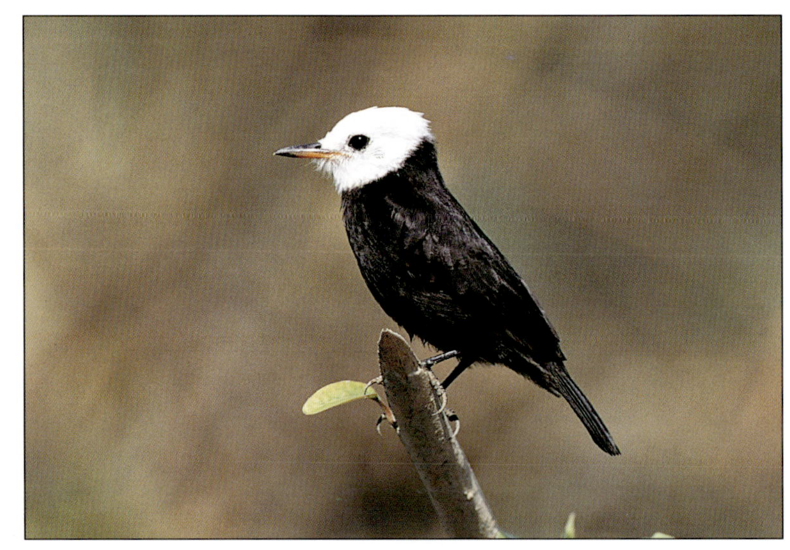

White-headed Marsh-Tyrant
Arundinicola leucocephala

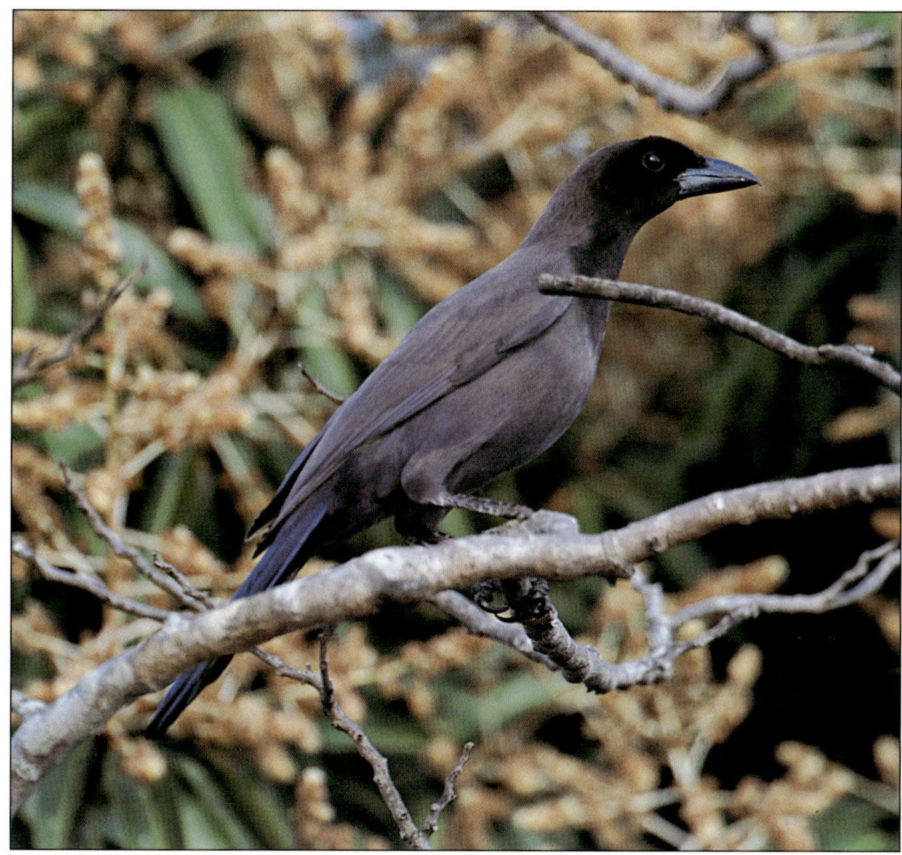

Purplish Jay
Cyanocorax cyanomelas

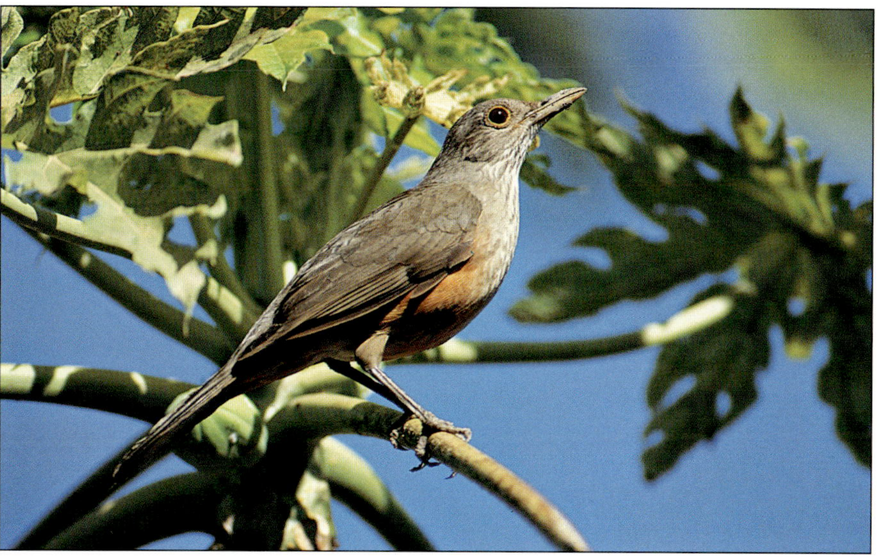

Rufous-bellied Thrush
Turdus rufiventris

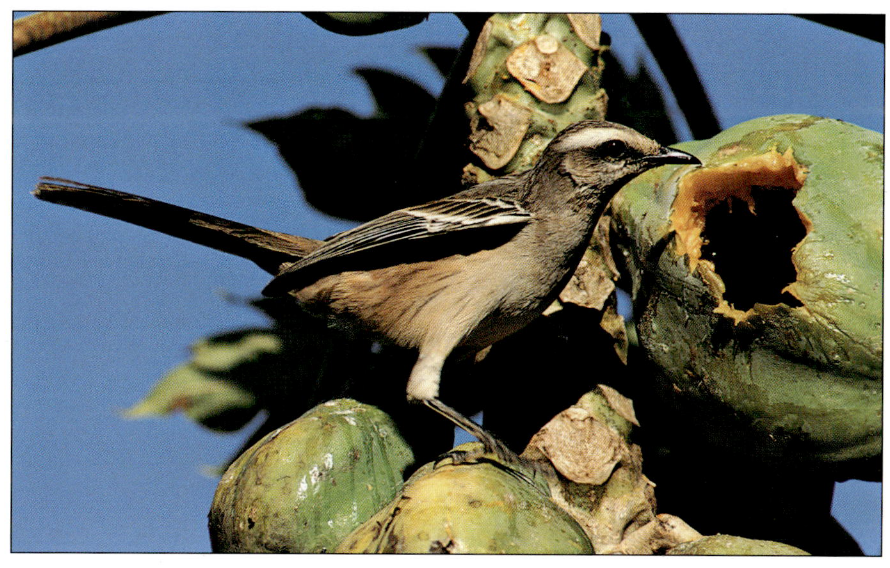

Chalk-browed Mockingbird
Mimus saturninus

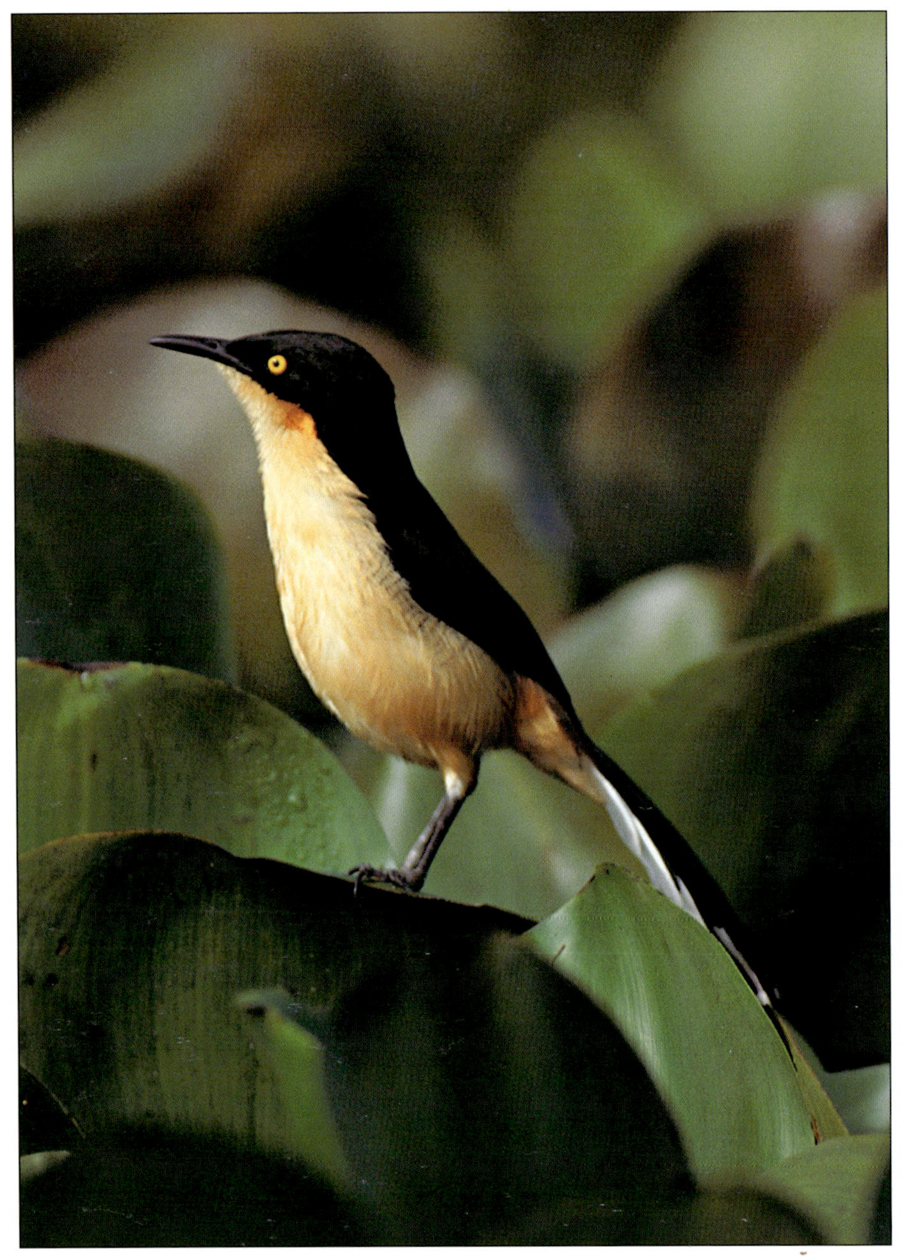

Black-capped Mocking-Thrush
Donacobius atricapillus

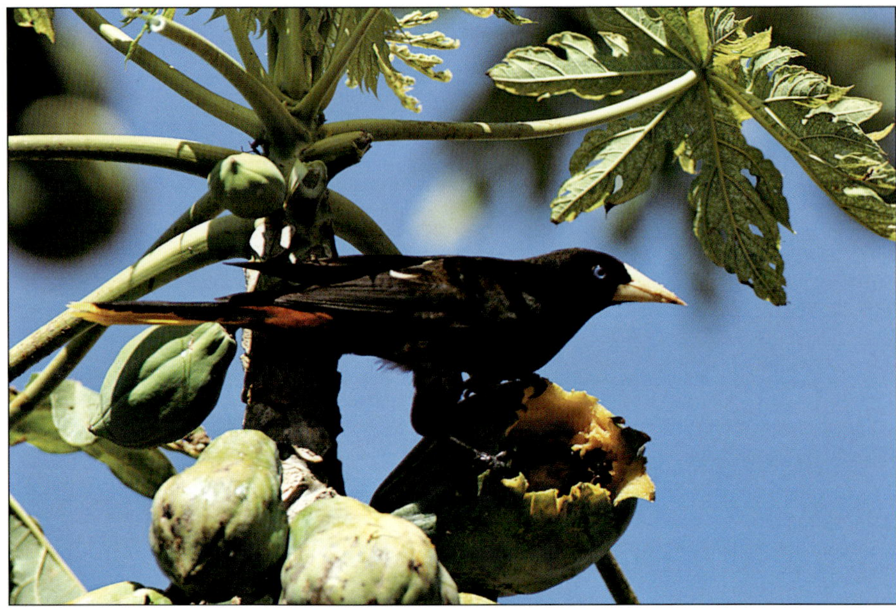

Crested Oropendola
Psarocolius decumanus

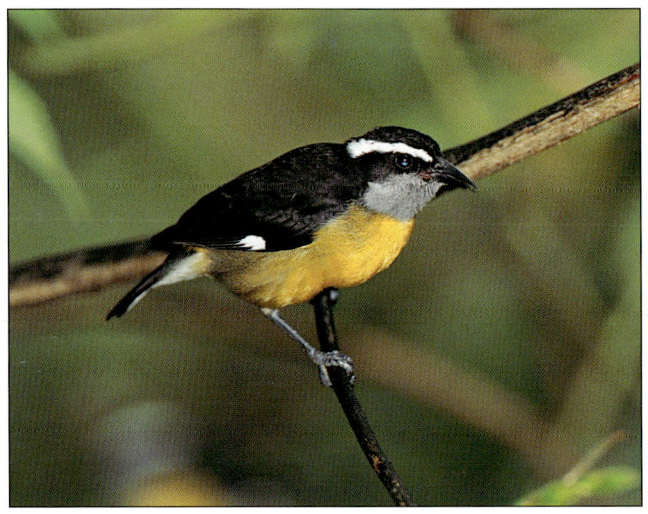

Bananaquit
Coereba flaveola

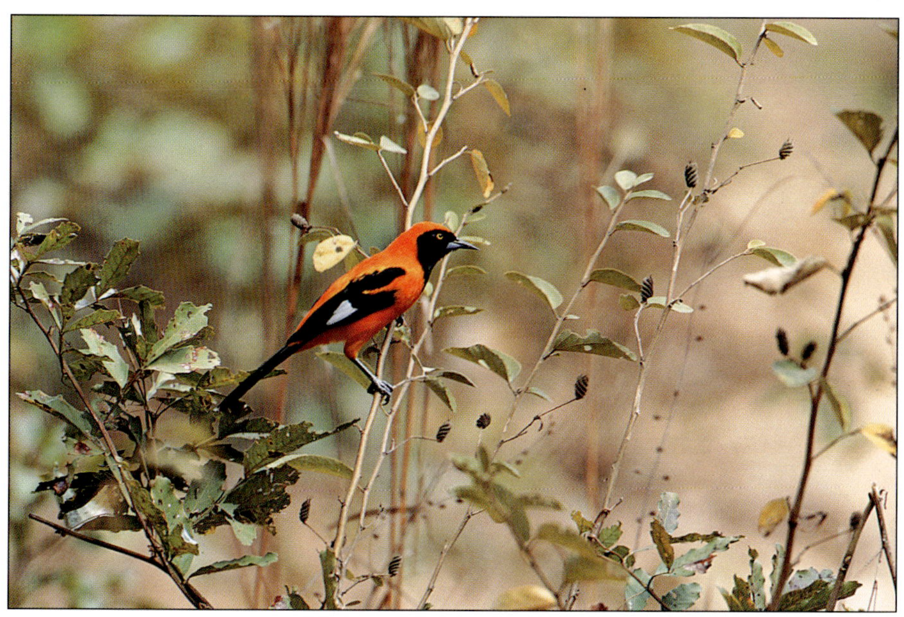

Troupial
Icterus icterus

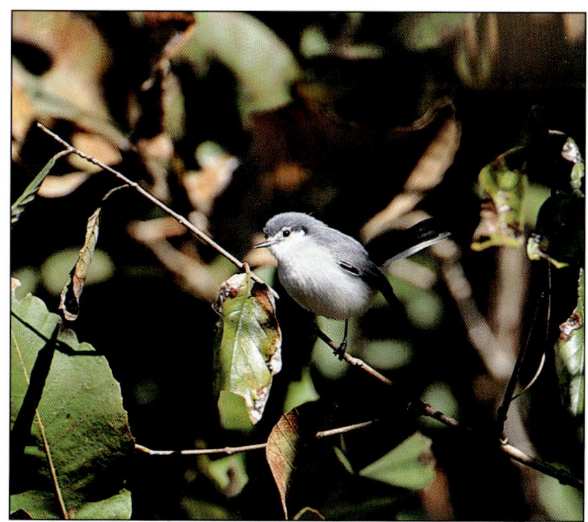

Cream-bellied Gnatcatcher
Polioptila lactea

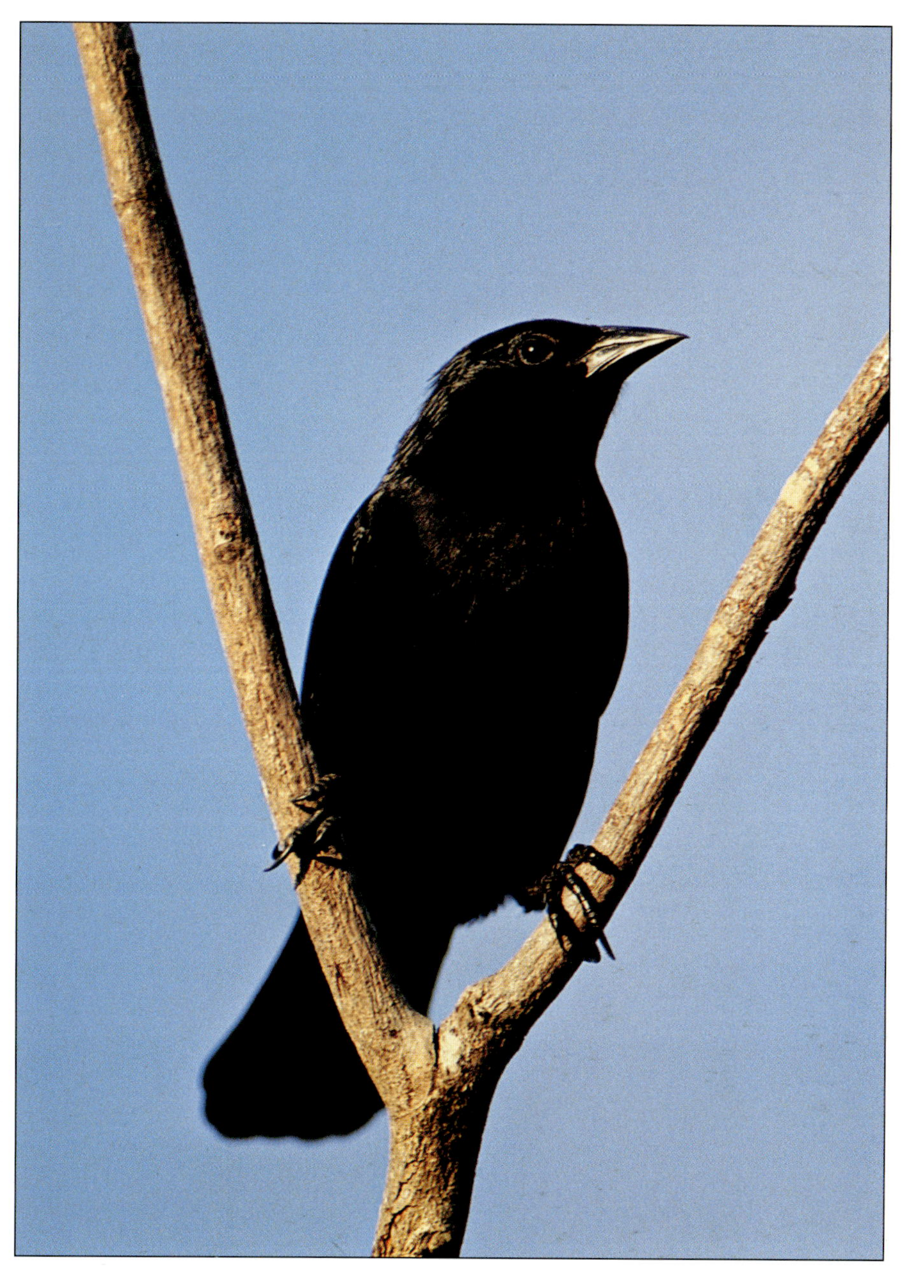

Chopi Blackbird
Gnorimopsar chopi

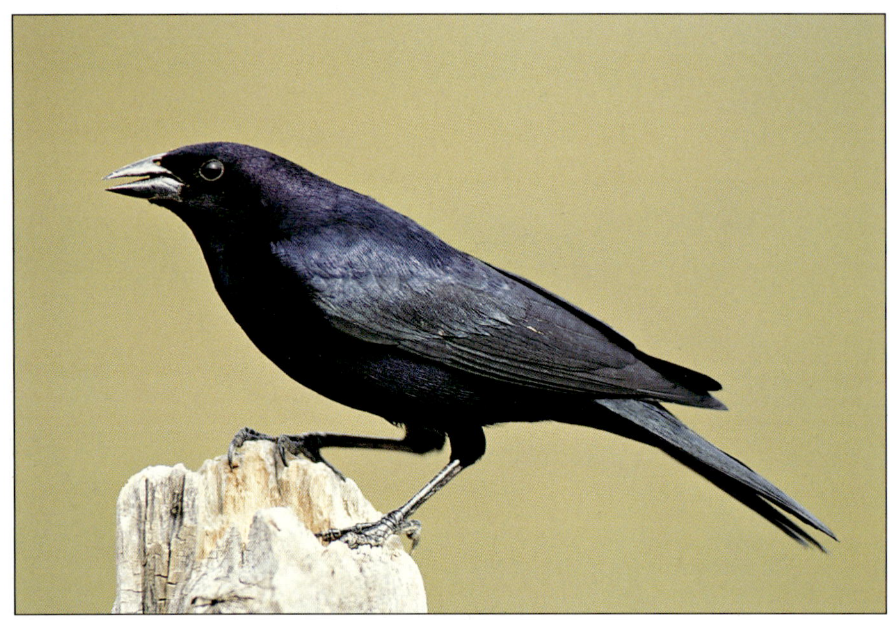

Shiny Cowbird
Molothrus bonariensis

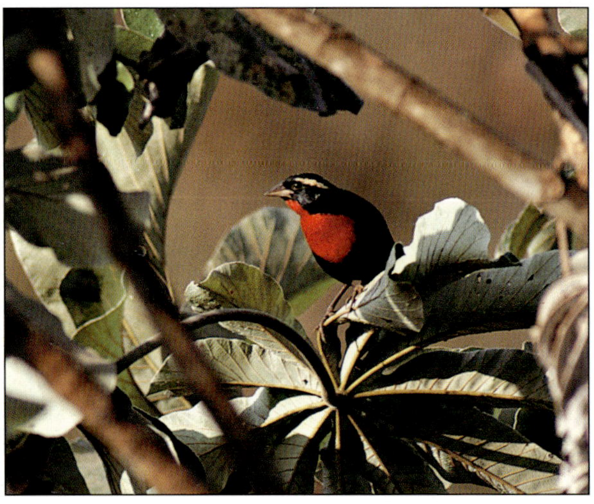

White-browed Blackbird
Leistes superciliaris

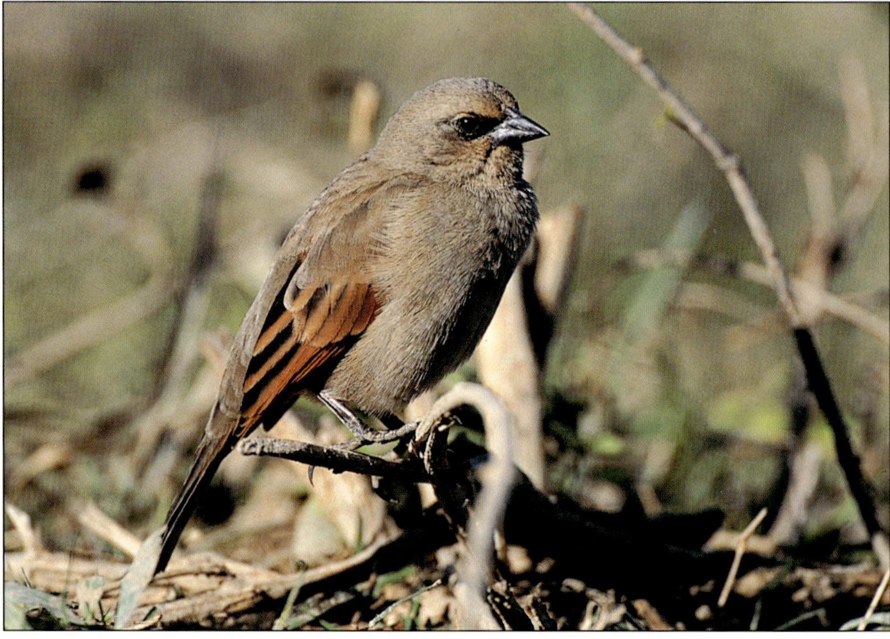

Bay-winged Cowbird
Molothrus badius

The Shiny Cowbird is often confused with the Chopi Blackbird. Look for distinctive grooves on the lower bill of the Chopi Blackbird. This is the best characteristic to separate them along with the Shiny Blackbird having a purplish gloss to it.

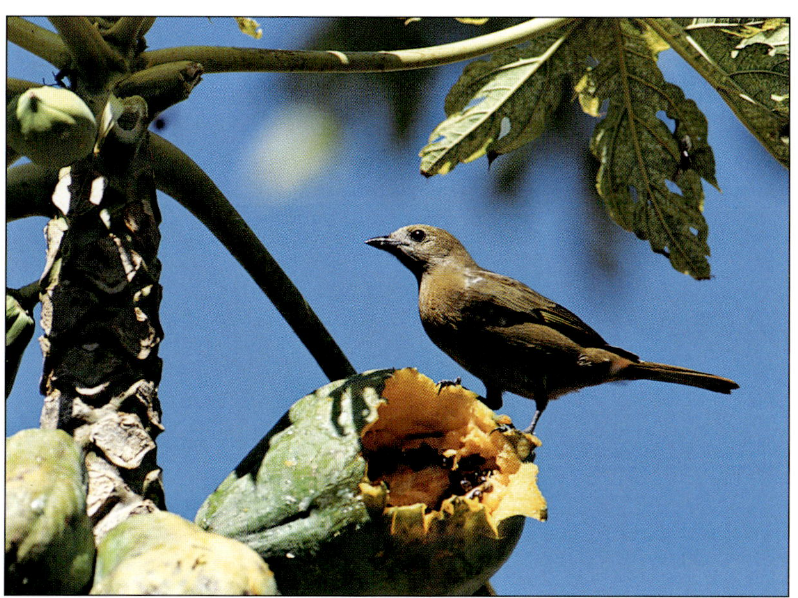

Palm Tanager
Thraupis palmarum

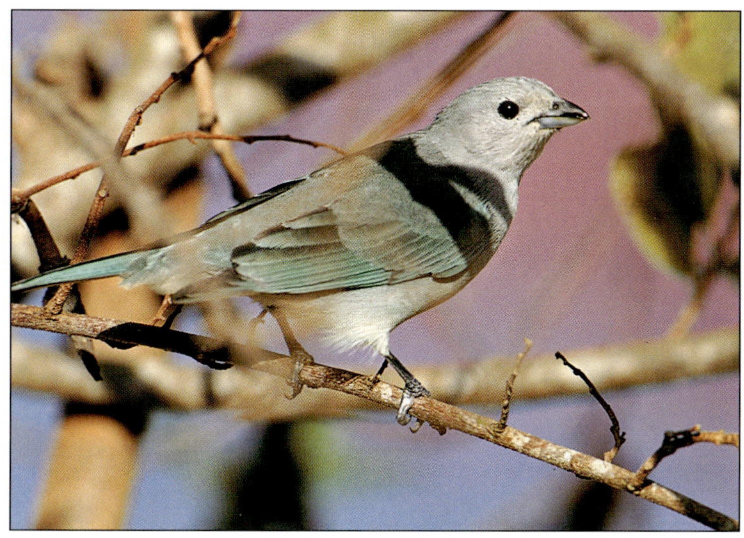

Sayaca Tanager
Thraupis sayaca

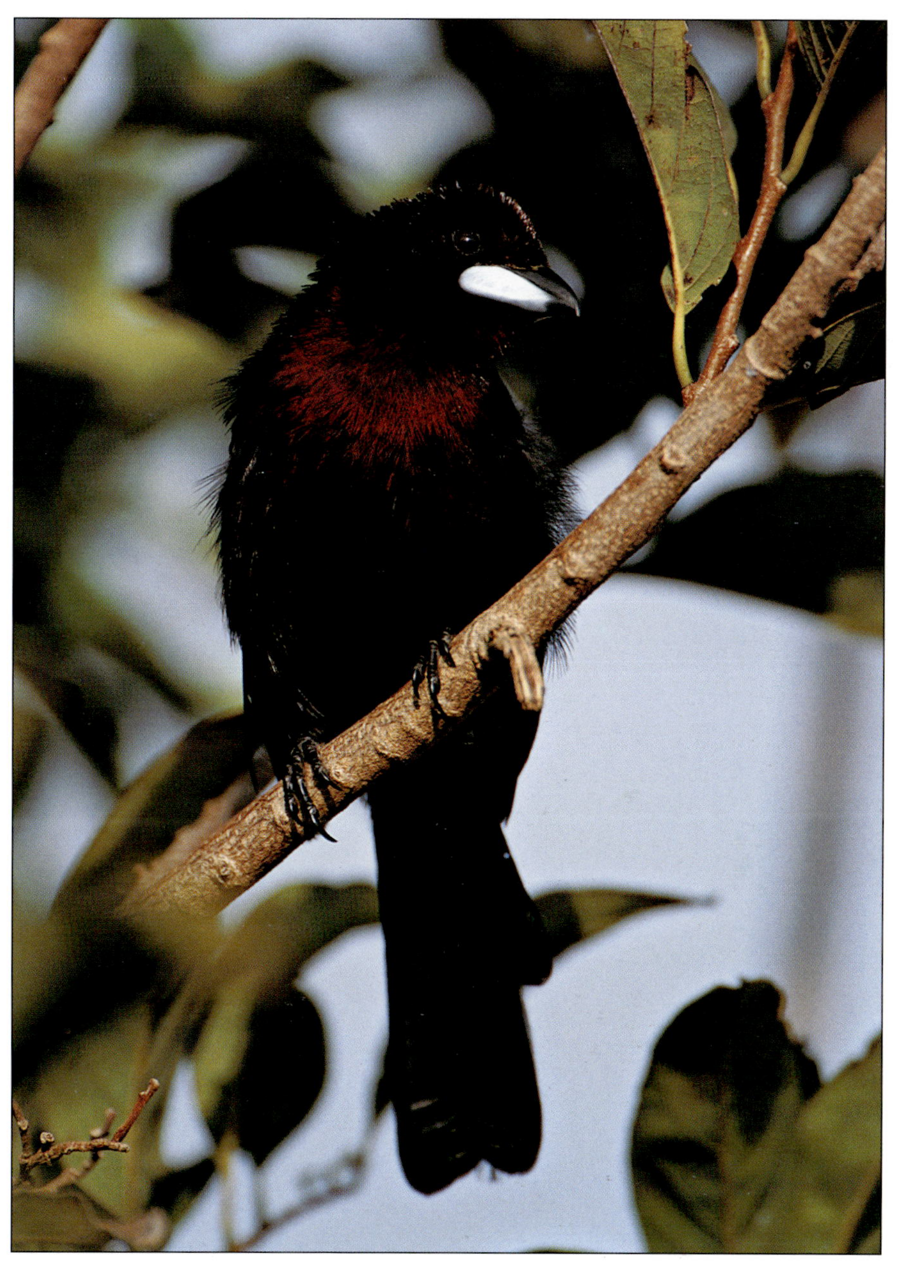

Silver-beaked Tanager
Ramphocelus carbo

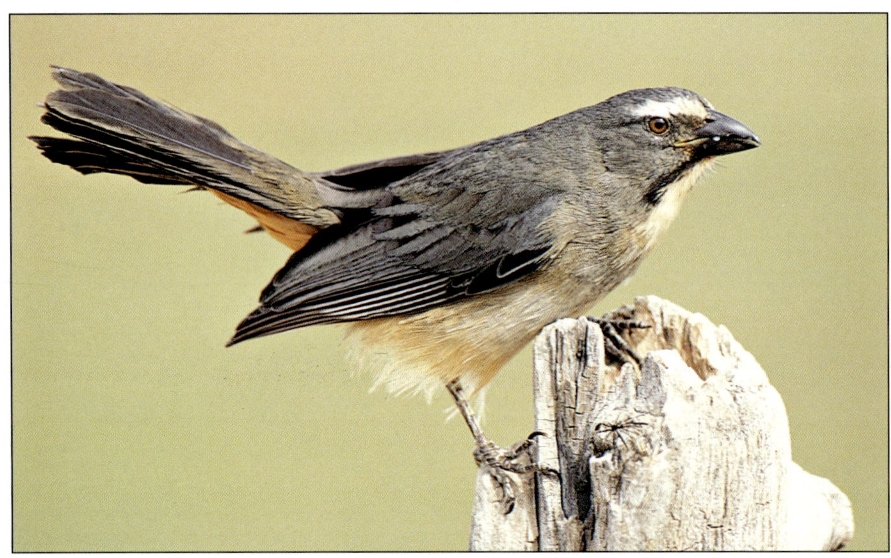

Grayish Saltator
Saltator coerulescens

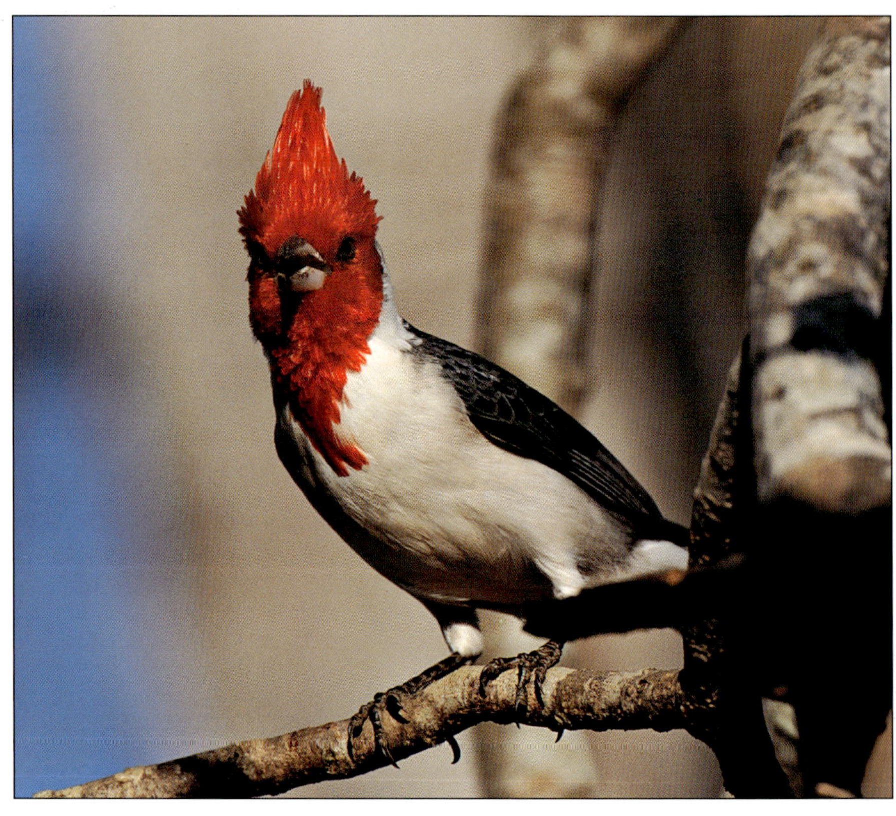

Red-crested Cardinal
Paroaria coronata

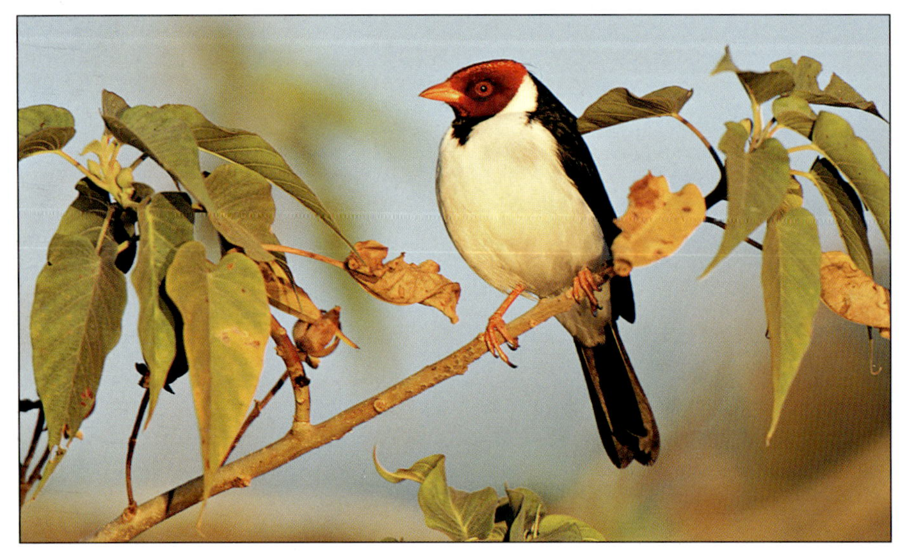

Yellow-billed Cardinal
Paroaria capitata

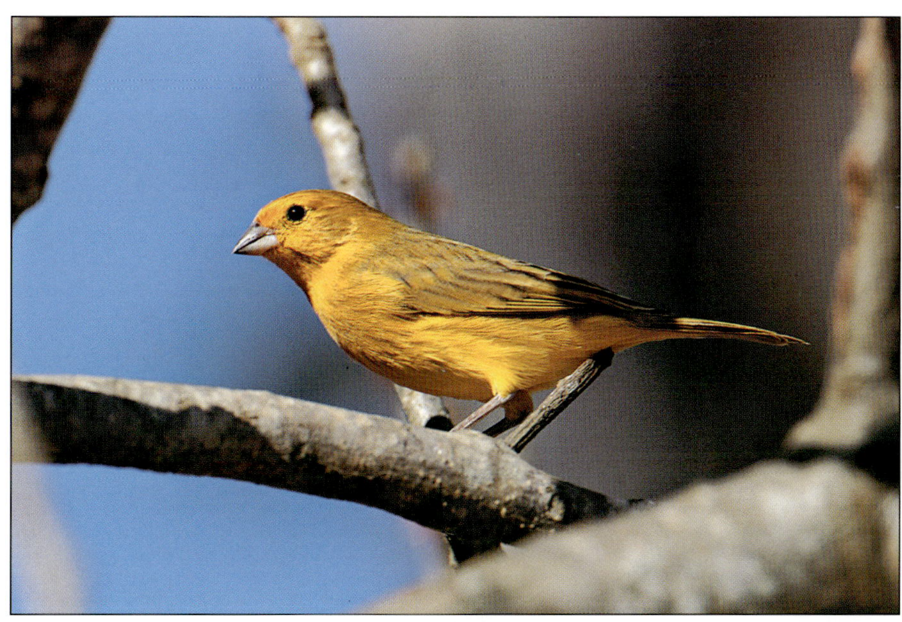

Saffron Yellow-Finch
Sicalis flaveola

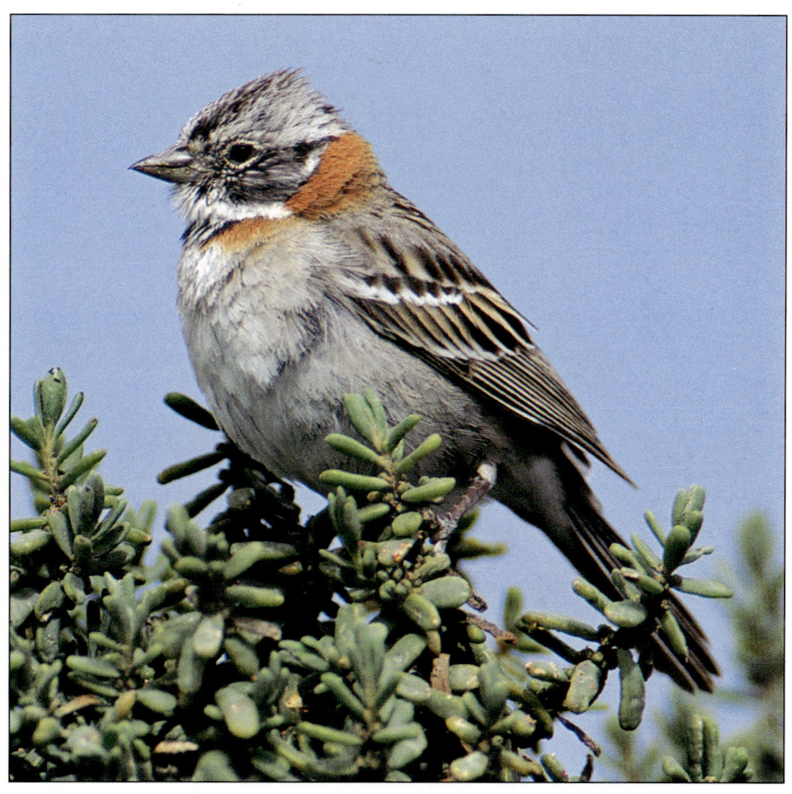

Rufous-collared Sparrow
Zonotrichia capensis

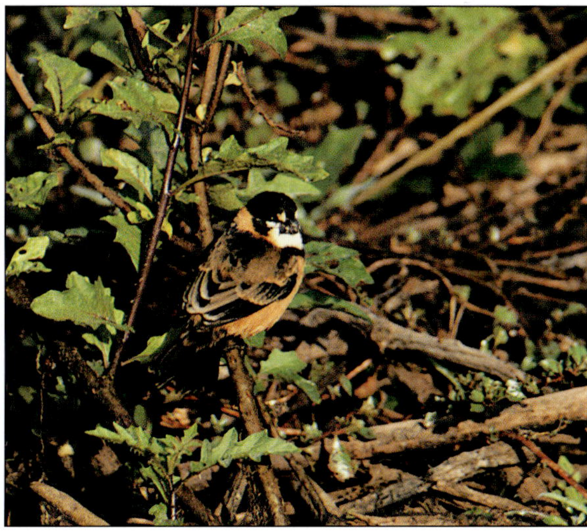

Rusty-collared Seedeater
Sporophila collaris

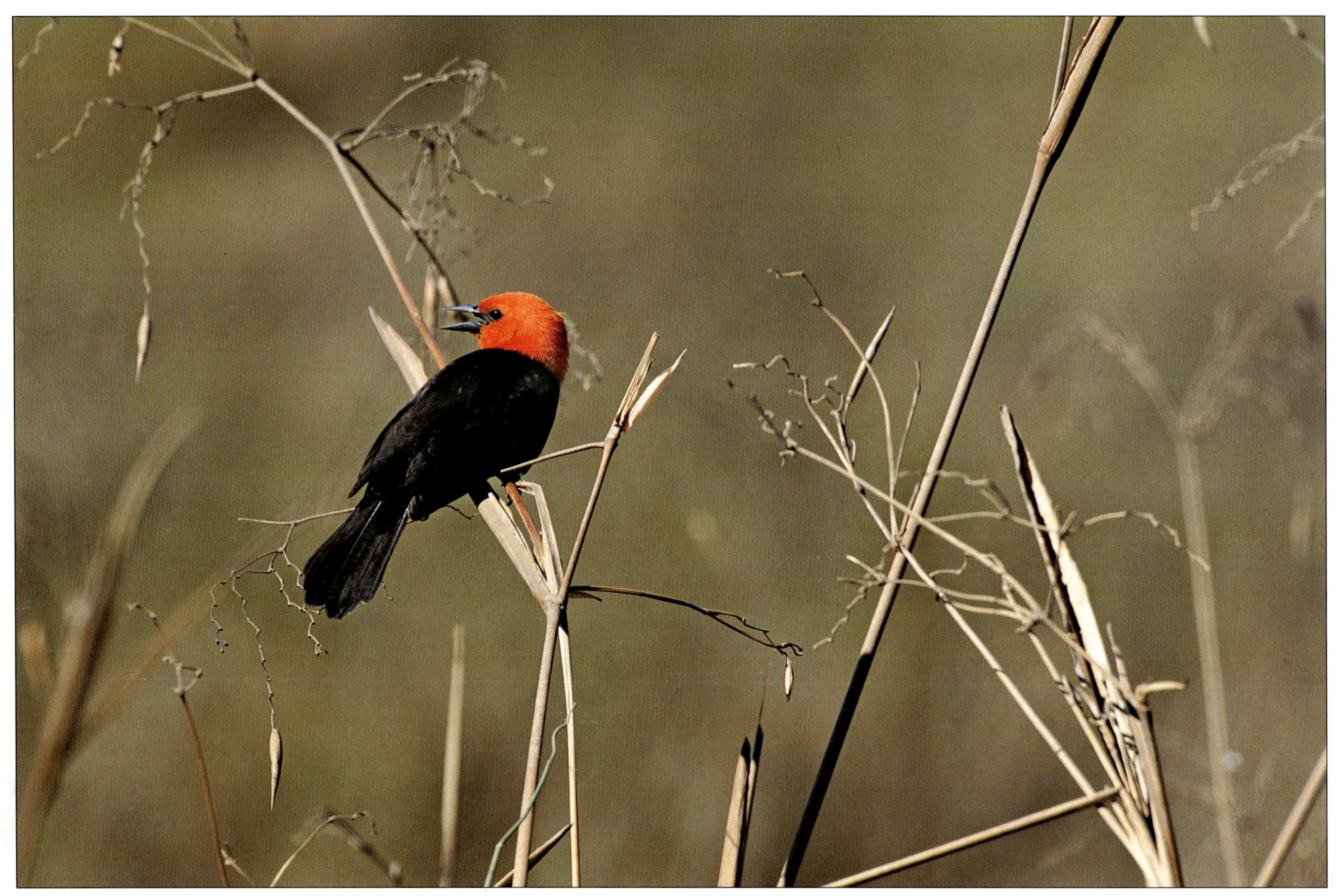

Scarlet-headed Blackbird
Amblyramphus holosericeus

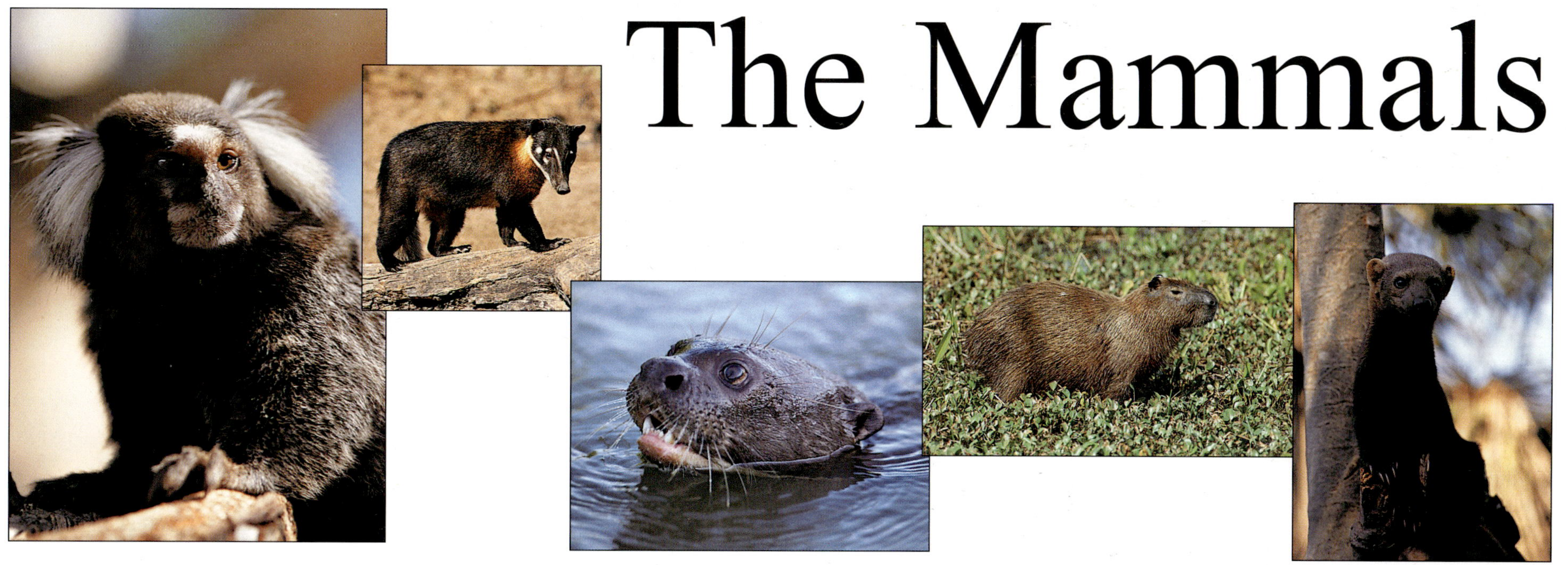

The Mammals

While the dense rainforest of the Amazon conceals most of its animals, the vast openness of the Pantanal is just the opposite. This is especially true with the mammals. Here they regularly present themselves in full view. Much of the time in front of surprised visitors with cameras or binoculars in hand. So far, scientistis have identified over 80 mammals and feel there are some that have not yet been categorized, which could bring the total higher.

Mammals have a little more difficult time adjusting to the dry/wet seasonal change. Unlike birds, they cannot fly off to more productive food resources. Mammals tied to the land adapt their ranges and food resources to areas of dry land that shrink and grow with the seasons. In the wet season many will seek shelter in higher elevations known as "capões" or wooded mounds. Some larger mammals migrate to higher land and may leave the Pantanal altogether.

Again, with such a rich ecological paradise, the Pantanal boasts some of the largest members in various groups. The largest deer, monkey and rodent in South America all reside here. They are the Pantanal or Marsh Deer, the Howler Monkey and the Capybara. The largest cat in the Americas, the Jaugar, still has a toe hold. The giant river otter, the largest otter in the world, can be found cruising up and down many of the rivers.

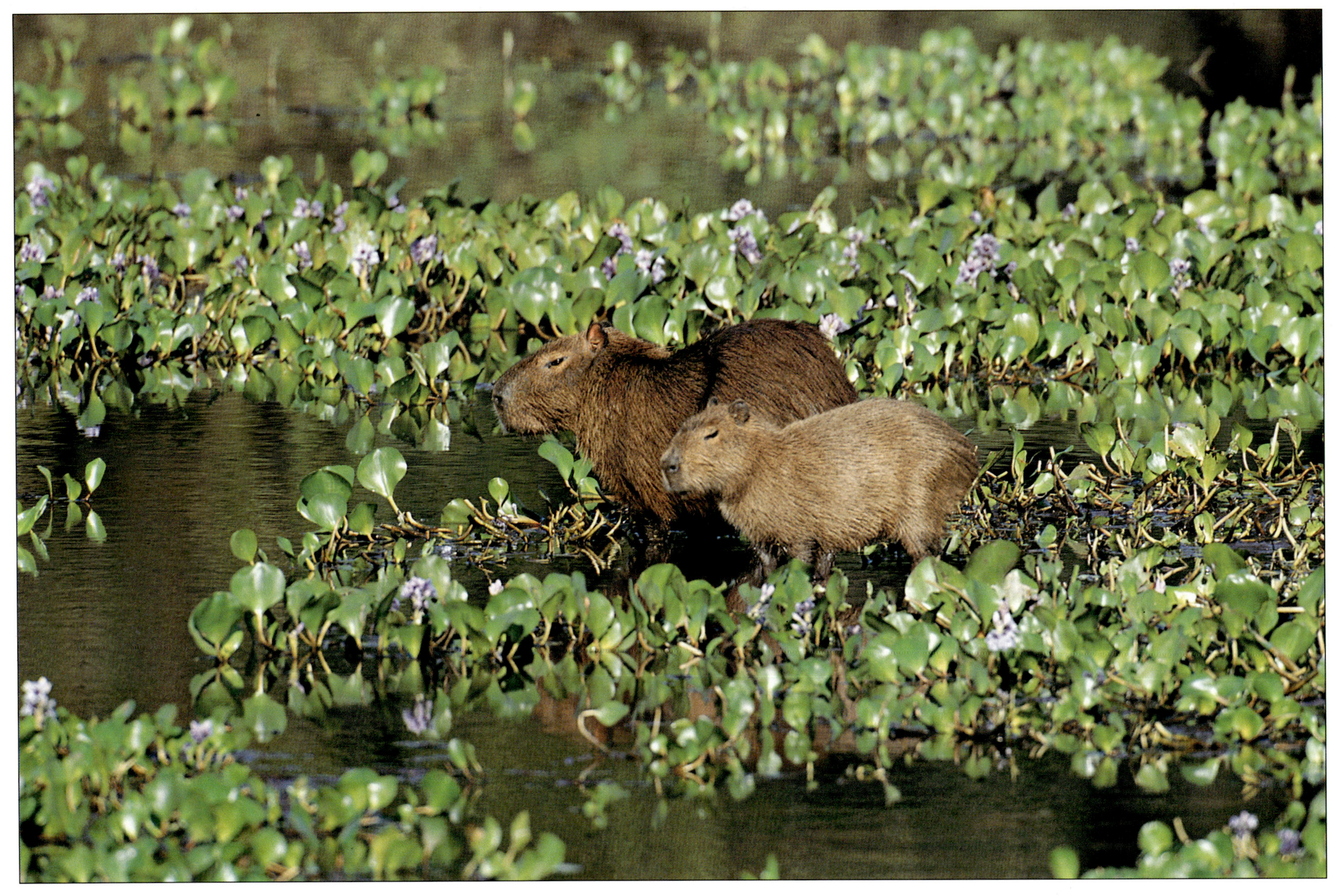

Capybara
Hydrochoerus hydrochaeris

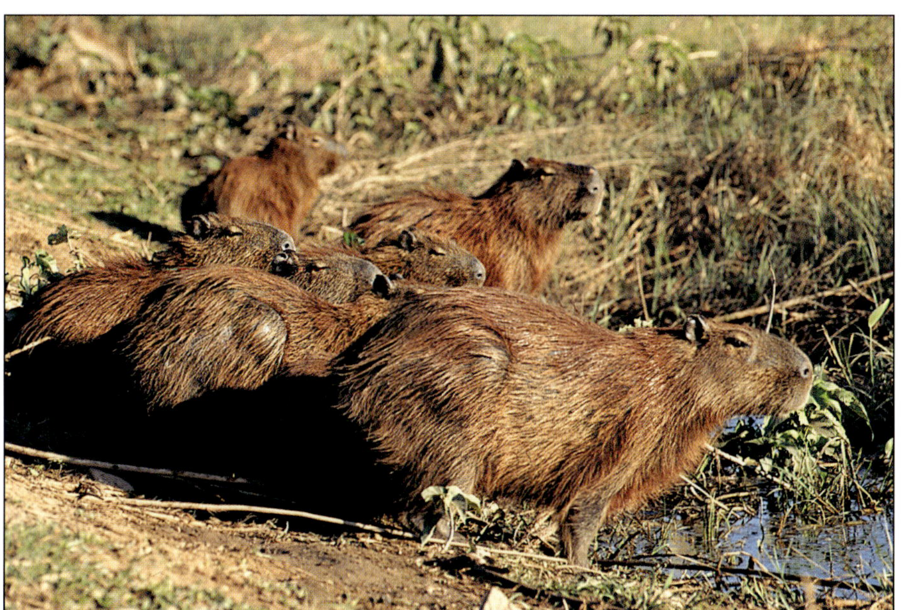

Social group of Capybaras

Weighing in at over 100 pounds, Capybara are the largest rodents in the world. Stocky, with short legs and ears, their most distinctive characteristic are webbed feet. This makes them semiaquatic and excellent swimmers. For escape, Capybara can either swim at the surface with eyes, ears and nostrils exposed or they can swim totally submerged. Swimming only as far as to find cover among the water lilies, they will remain there until danger passes. It is not uncommon to find groups of about 20 Capybara because they are extremely social animals. When feeding, they graze using incisors to clip vegetation at ground level. With extensive chewing and hindgut fermentation, they can digest what is eaten with great efficiency.

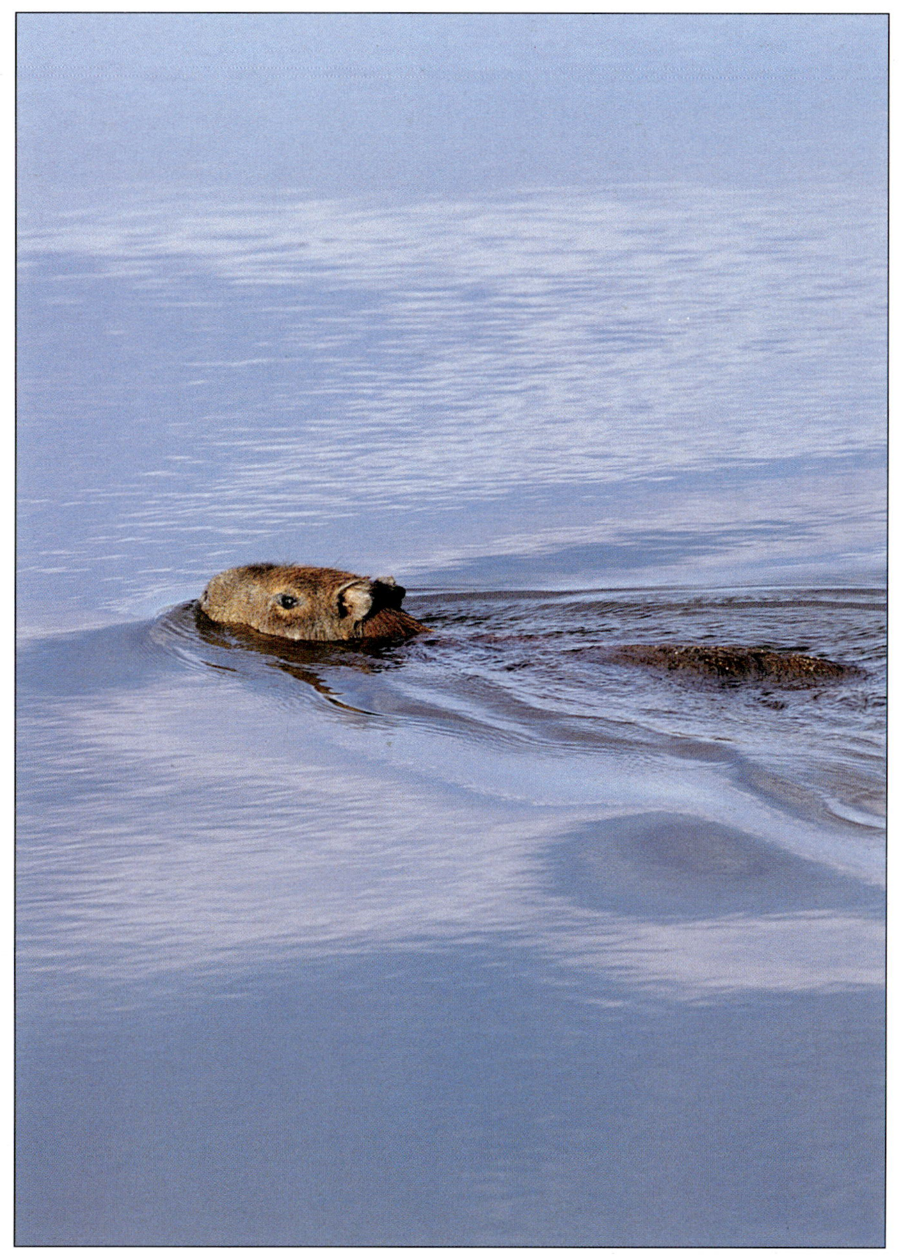

Capybara are aquatic and take to water readily.

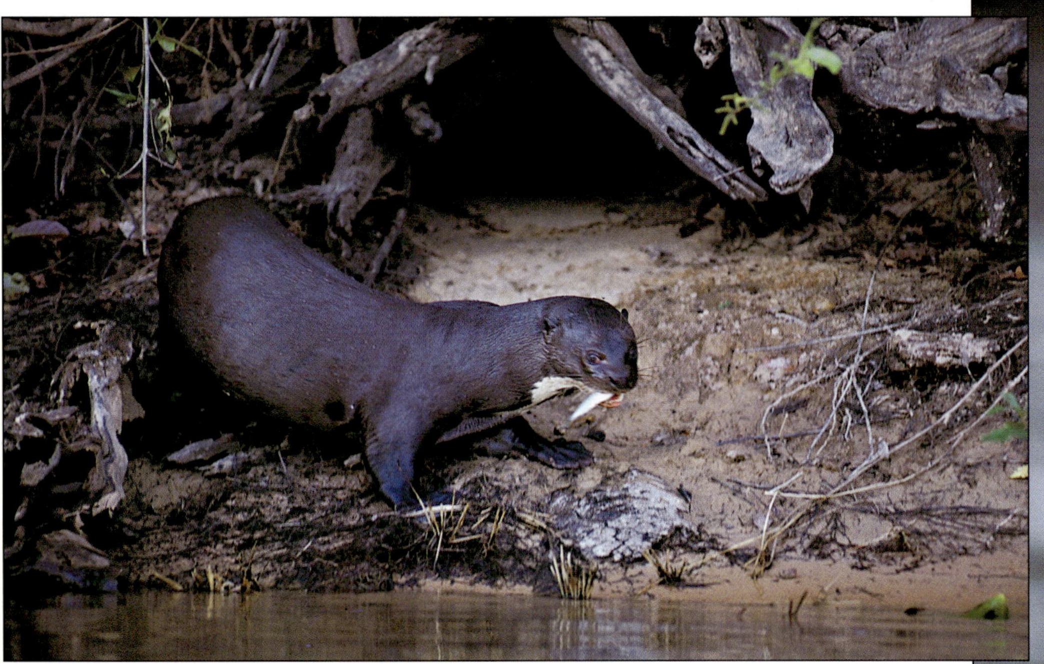

Giant River Otter at the entrance
of an old den site.

Giant River Otters have a precarious foothold in the Pantanal. Up until 1970 they were heavily sought after in the fur trade. Today, even though protected, otters still might come under the gun from fishermen feeling they compete for the same fish. As the name implies, they are the largest otter in the world attaining a length of six feet. Strict carnivores, they eat anything from fish to caiman to snakes. If one is observed, several more of these social critters are usually nearby.

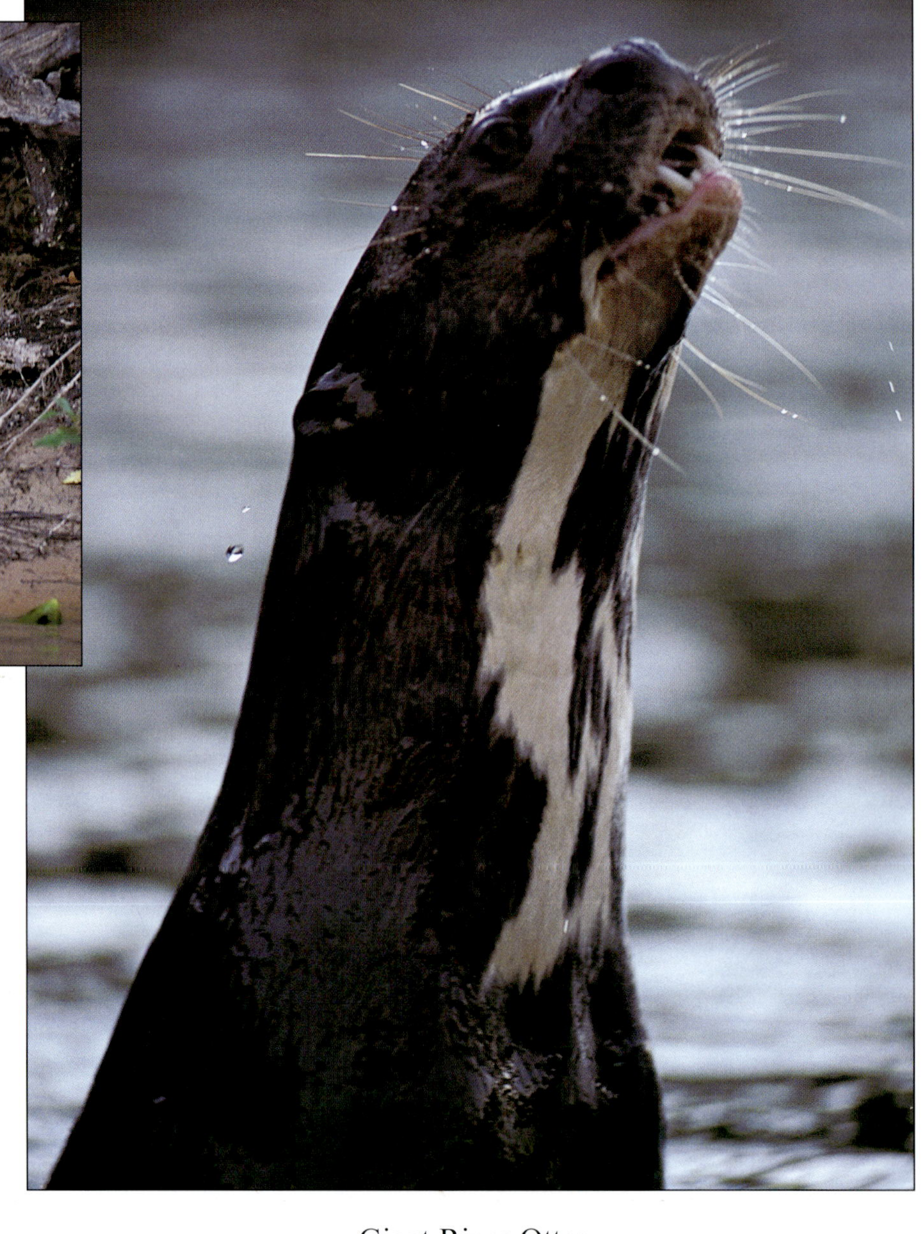

Giant River Otter
Pteronura brasiliensis

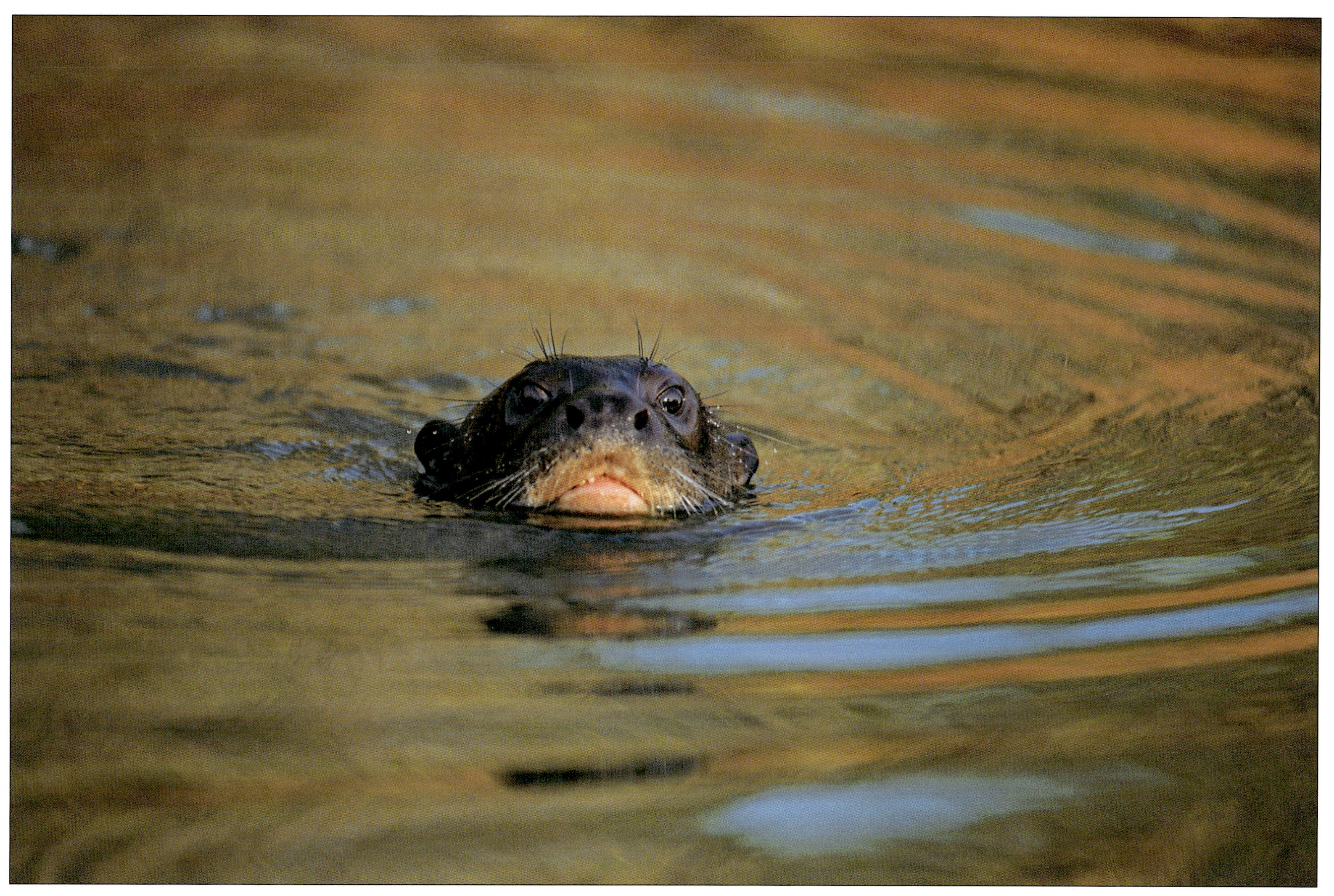

Giant River Otter swimming

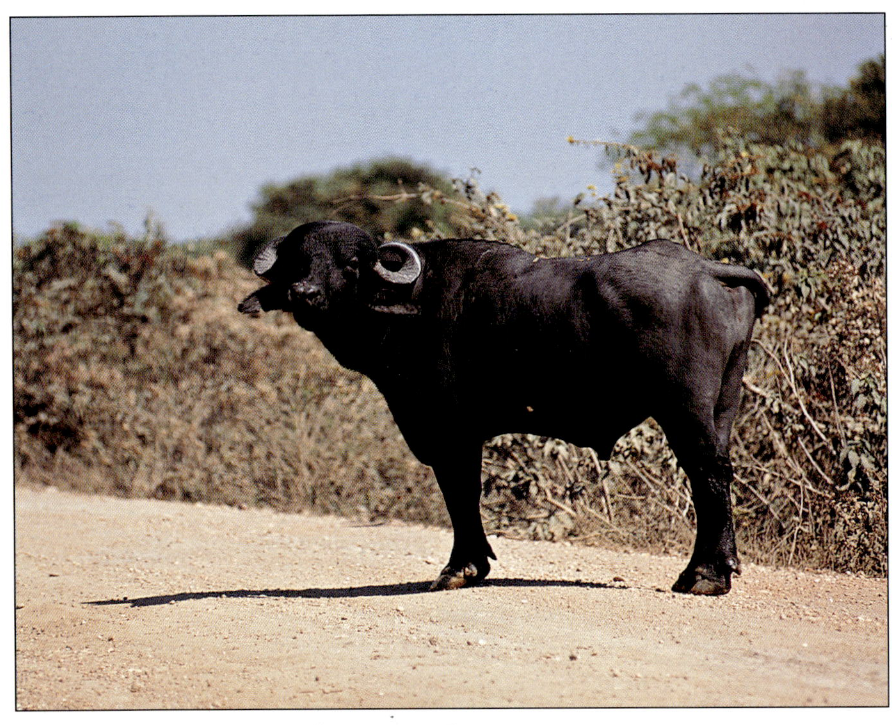

Water Buffalo

One recently introduced species that has adapted well is the Water Buffalo. It was introduced as another meat source for the growing world market. Since few fences can hold them back, they are free ranging, which puts stress on the environment. A native species feeling this pressure is the South American Marsh Deer. They once thrived in the tall grass of the flood plain. But competing with Water Buffalo, range cattle and the diseases they bring now threatens their existence. One factor in the deer's favor is that humans find their meat unpalatable.

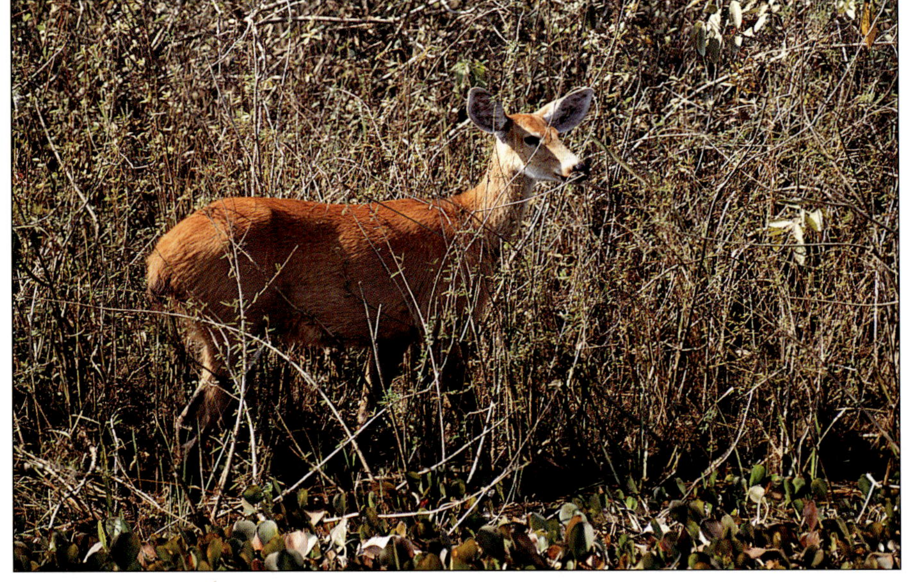

Pantanal or Marsh Deer
Blastocerus dichotomus

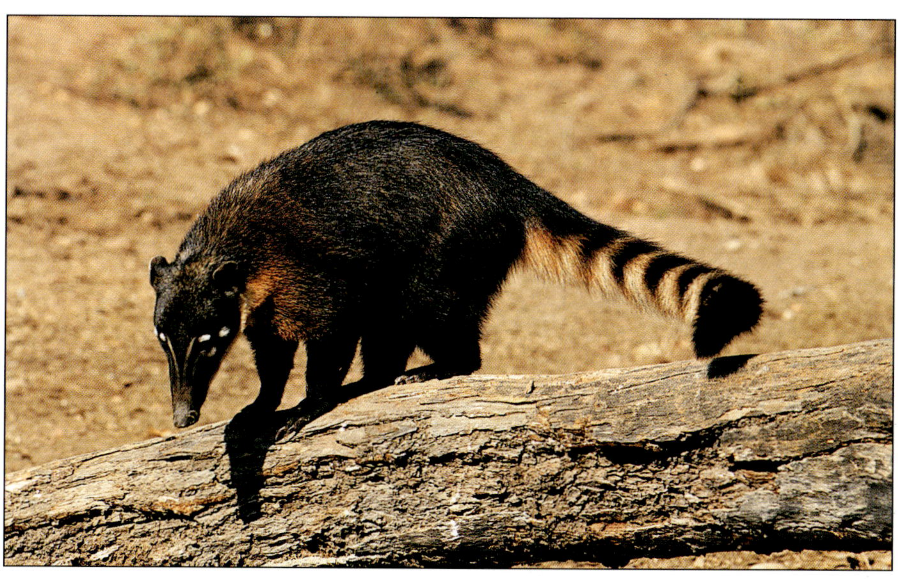

South American Coati
Nasua nasua

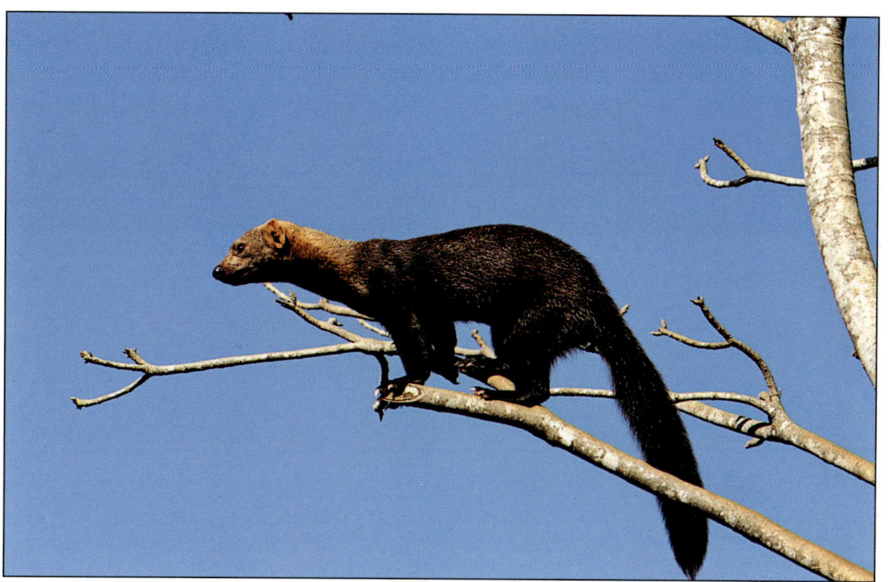

Tayra
Eira barbara

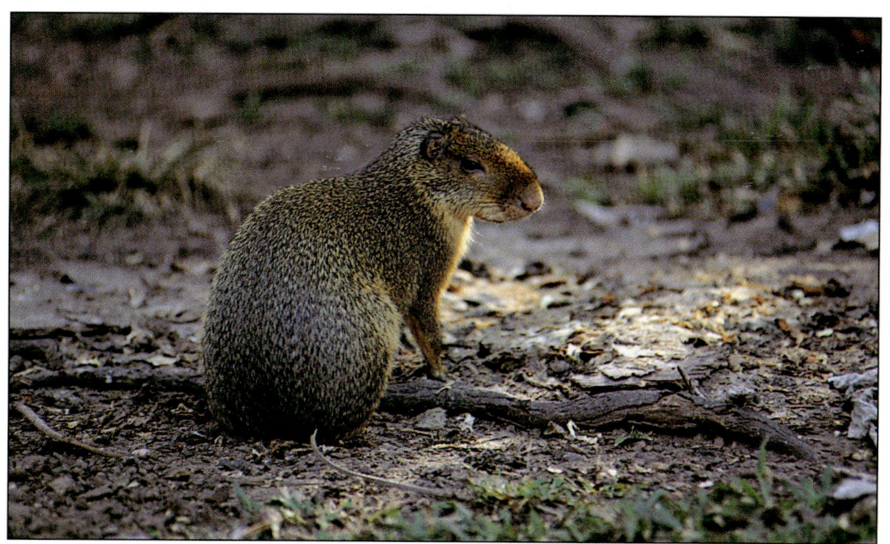

Agouti
Agoutie paca

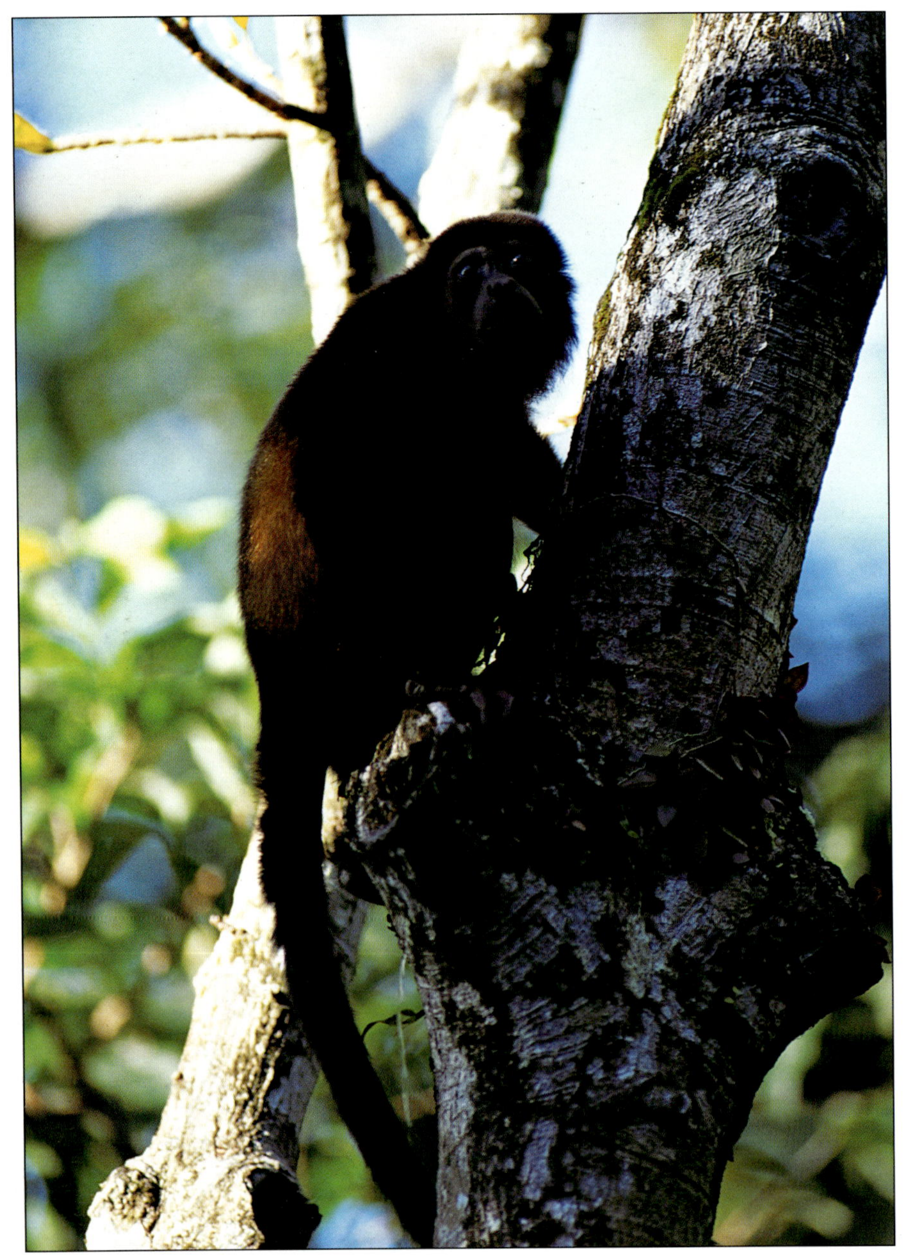

Howler Monkey
Alouatta caraya

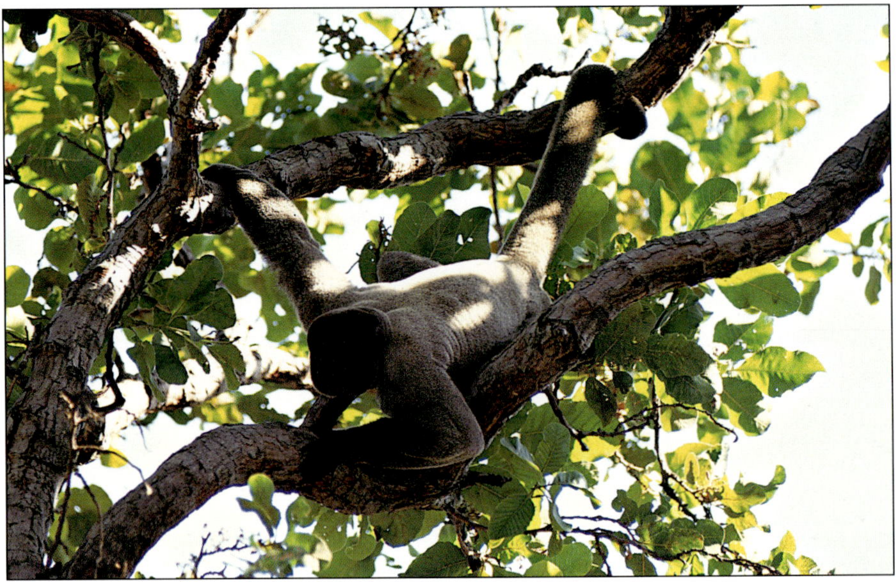

Woolly Monkey
Lagothrix lagothricha

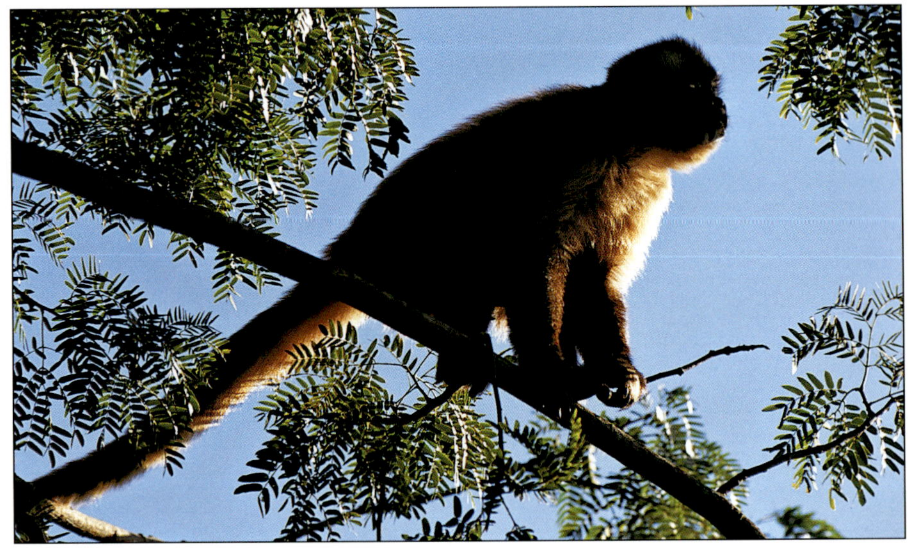

Brown Capuchin
Cebus apella

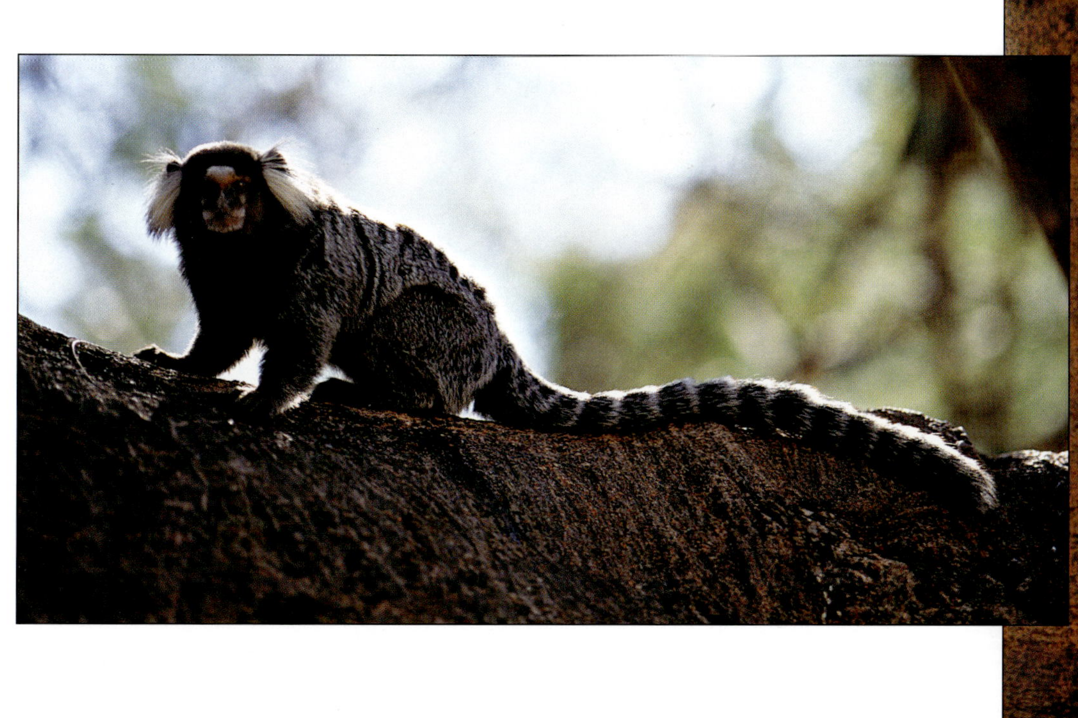

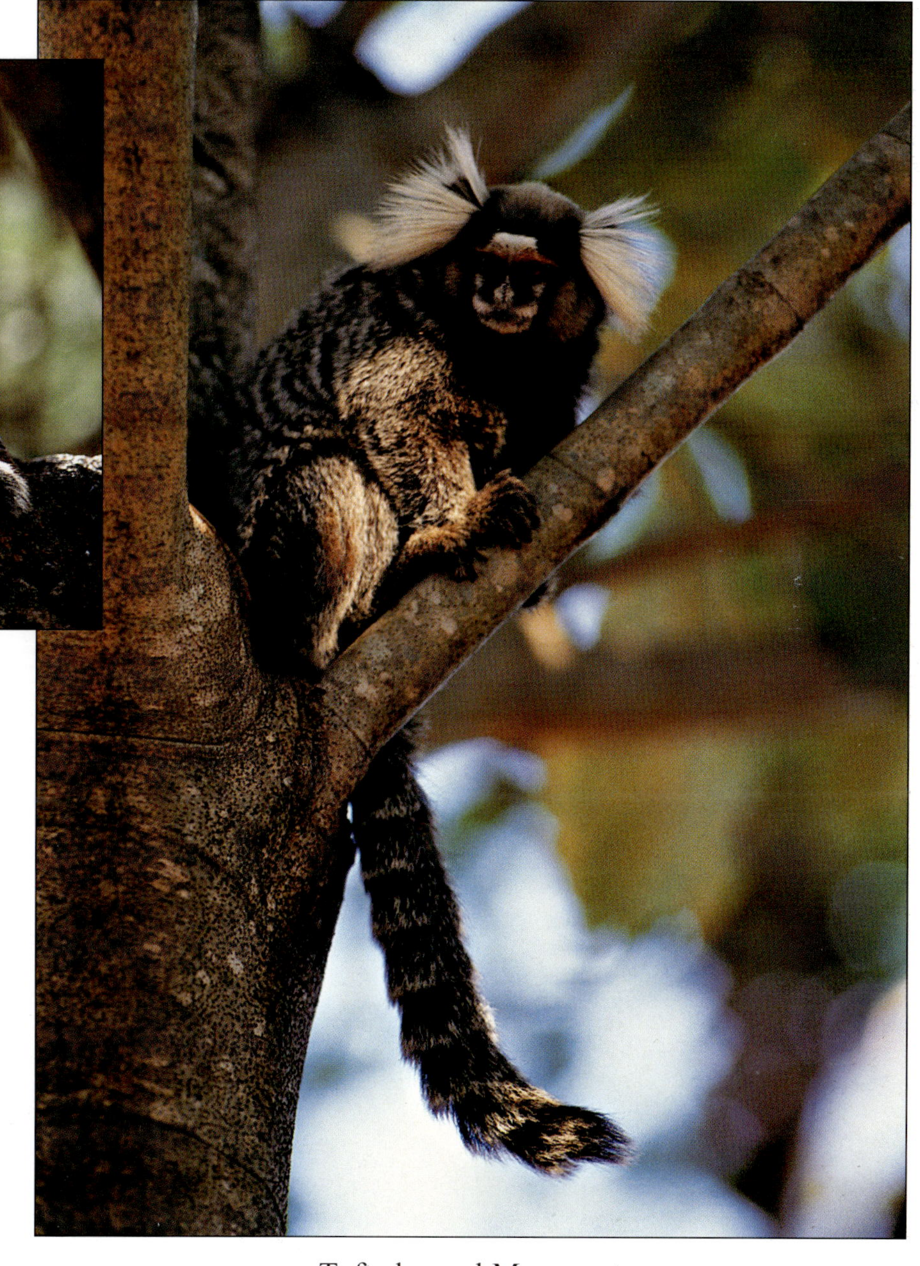

Primates fill a small niche in the whole ecological system of the Pantanal, because of the extensive clearing of trees when cattle ranching began in the 1800's. Large tracts of forest remain, but are so scattered it makes viewing any of the primates difficult. The best opportunities are in the forested corridors along rivers where troops of capuchins or howler monkeys can be seen on the move, foraging for buds, leaves, blooms and natural fruits. They move freely among the branches with the help of their prehensile tails that actually act as a fifth hand. For most of these species the end of the tail has a naked pad for better grip. The only reason they venture down from the canopy is to drink.

Tufted-eared Marmoset
Callithrix jacchus

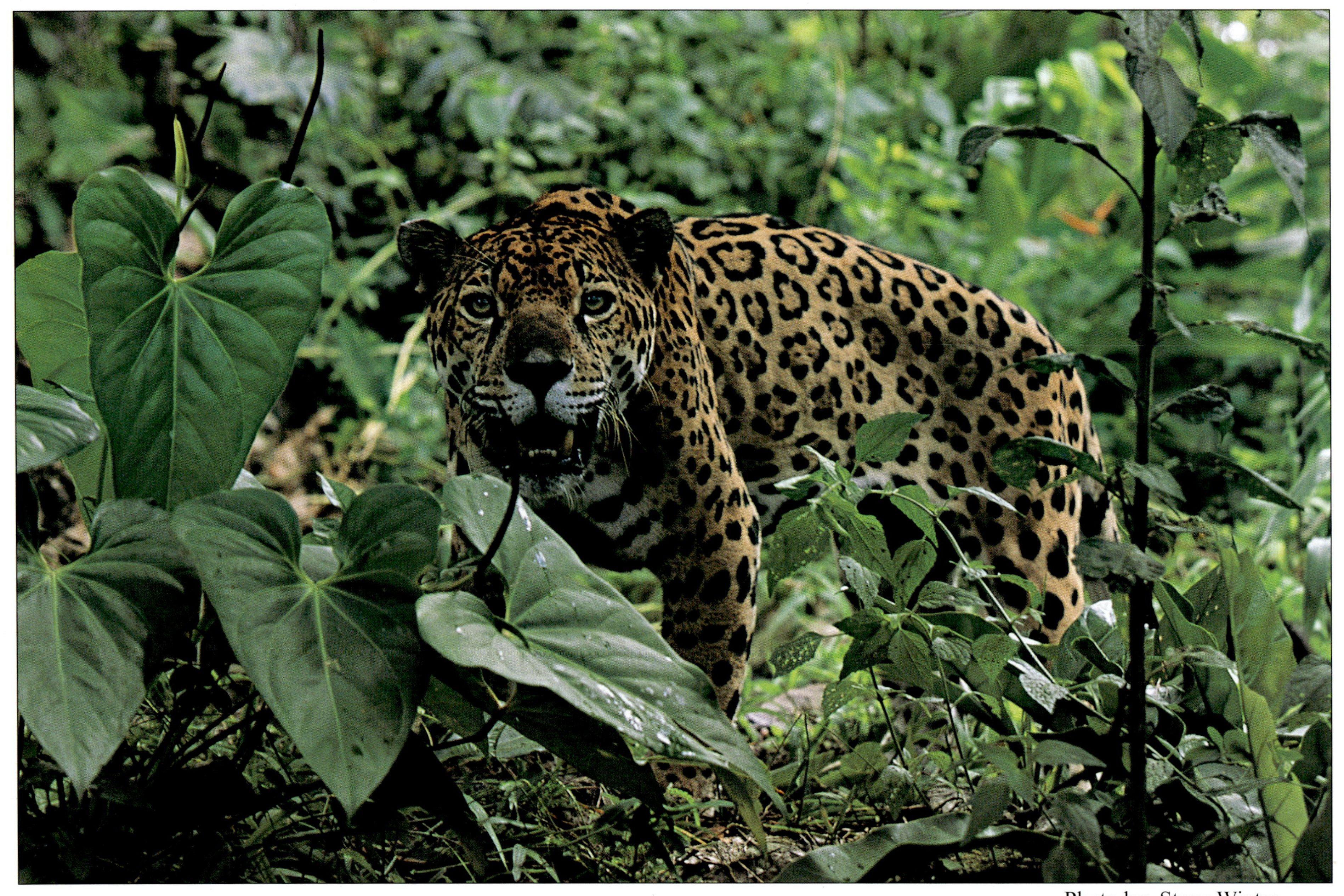

Photo by Steve Winter

Jaguar
Panthera onca

Everything else

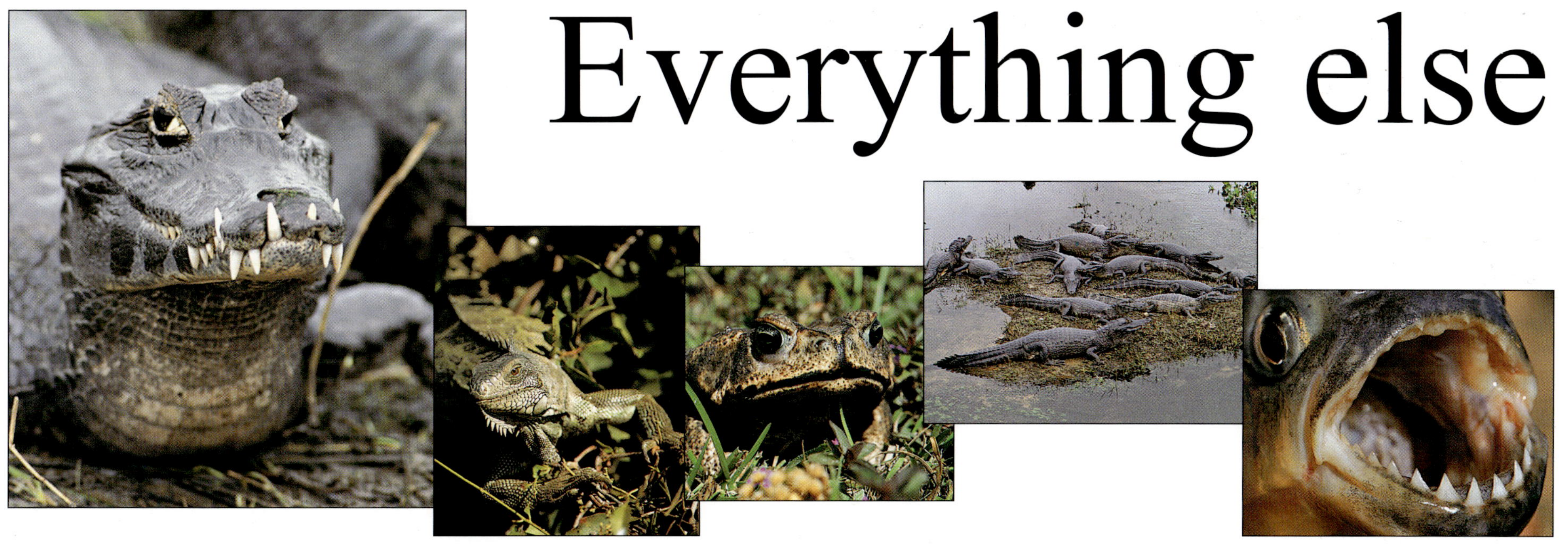

There are an estimated 50 different reptiles, 40 amphibians and 263 species of fish inhabiting the Pantanal. Except for a few members, most of these animals are a few links down from the top of the food chain. So when most often viewed, they are being eaten by birds or mammals. Therefore, to keep from becoming fill for some predator's stomach, cold blooded reptiles and amphibians spend most of their time motionless. This serves two purposes. It keeps them from being seen. It also allows them to bask in the sun to raise their body temperature.

The largest of the reptiles is the anaconda, which can grow to a length of 20 feet. Yet it is seldom seen. They are aquatic and will use the water as cover to lie in ambush waiting for prey to come within range. Caiman behave similarly. By doing this they are able to save considerable energy.

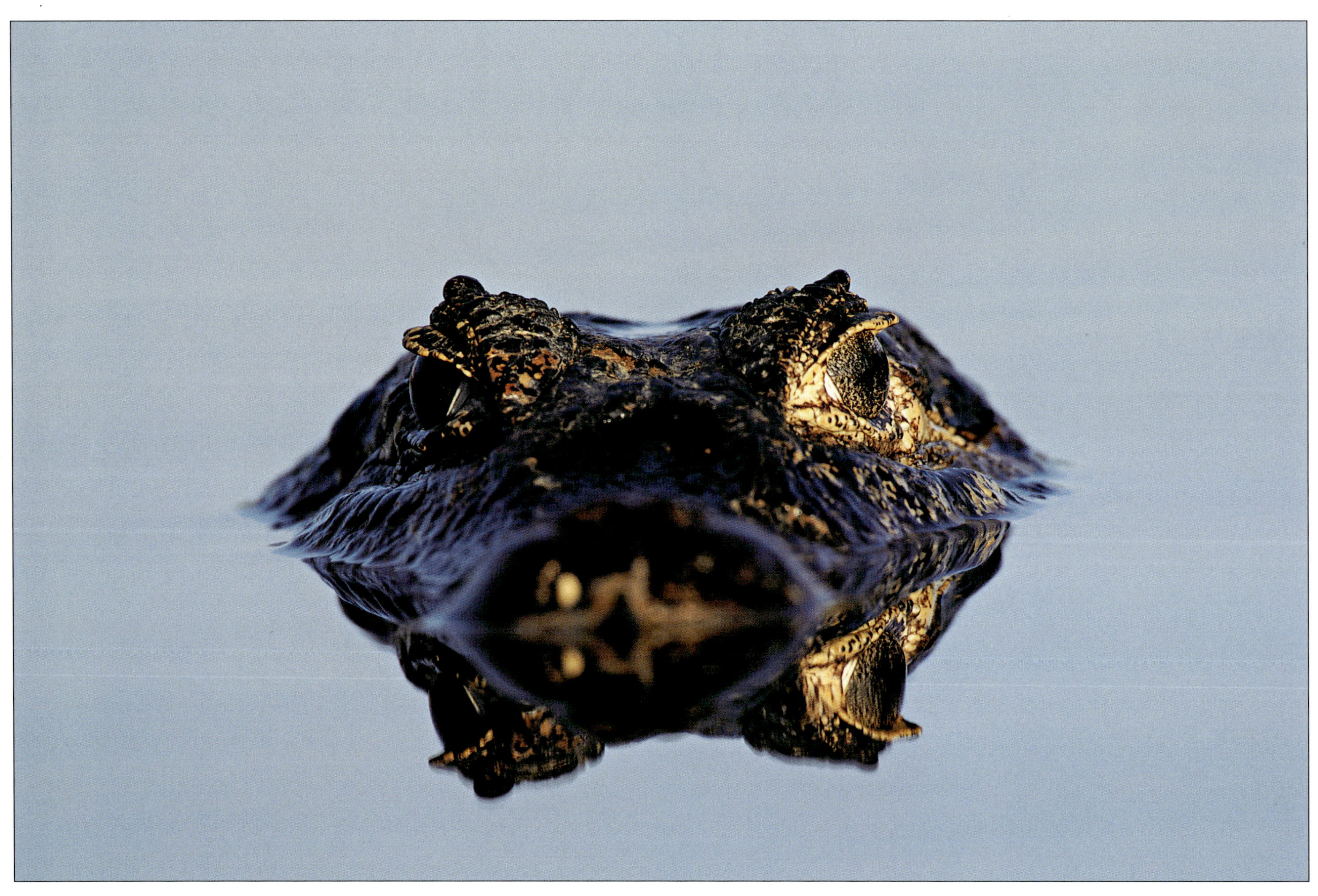

Caiman
Cayman jacare

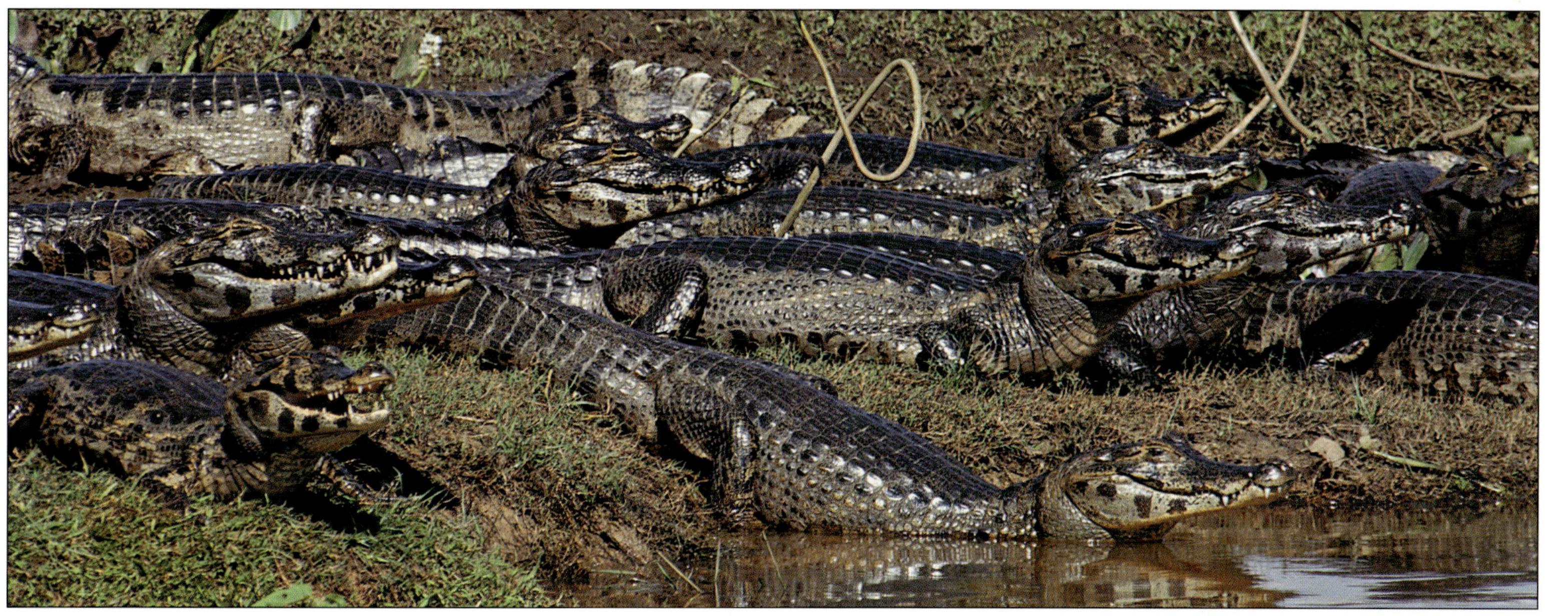

Resembling an alligator and often called "spectacled alligator," Caiman differ by a few physical characteristics. One is that they only attain a length of six or seven feet when fully mature. The main difference, which separates them, has to do with their eye orbits. With Caiman, the orbits are connected together across the skull and a prominent bony ridge surrounds each orbit. Extremely common in the Pantanal, Caiman become particularly noticeable in September toward the end of the dry season when receding water drives them into confined quarters. On land, they are timid and will flee when frightened. In the water, they are totally at home feeding on whatever the opportunity presents. Worn out or lost teeth are soon replaced by shiny new ones. One thing about Caiman is that they are a strong tourist attraction.

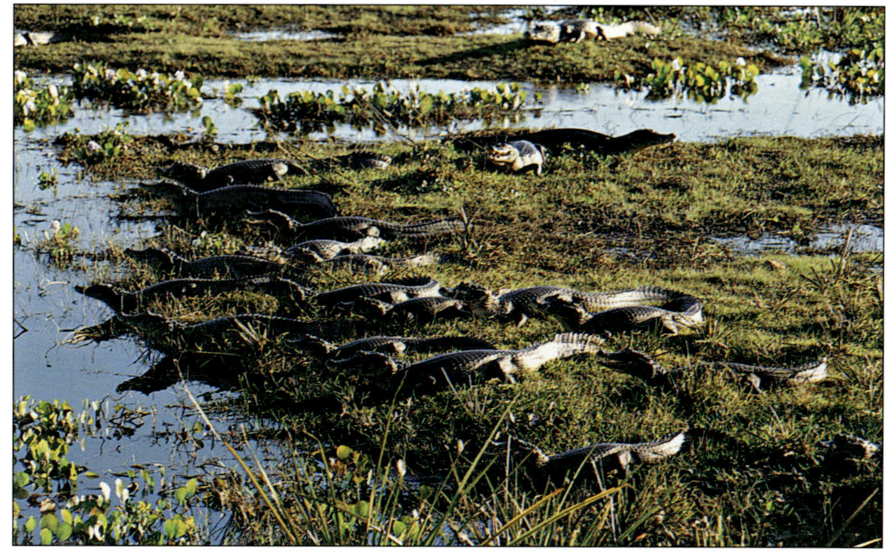

Caiman love to bask in the sun

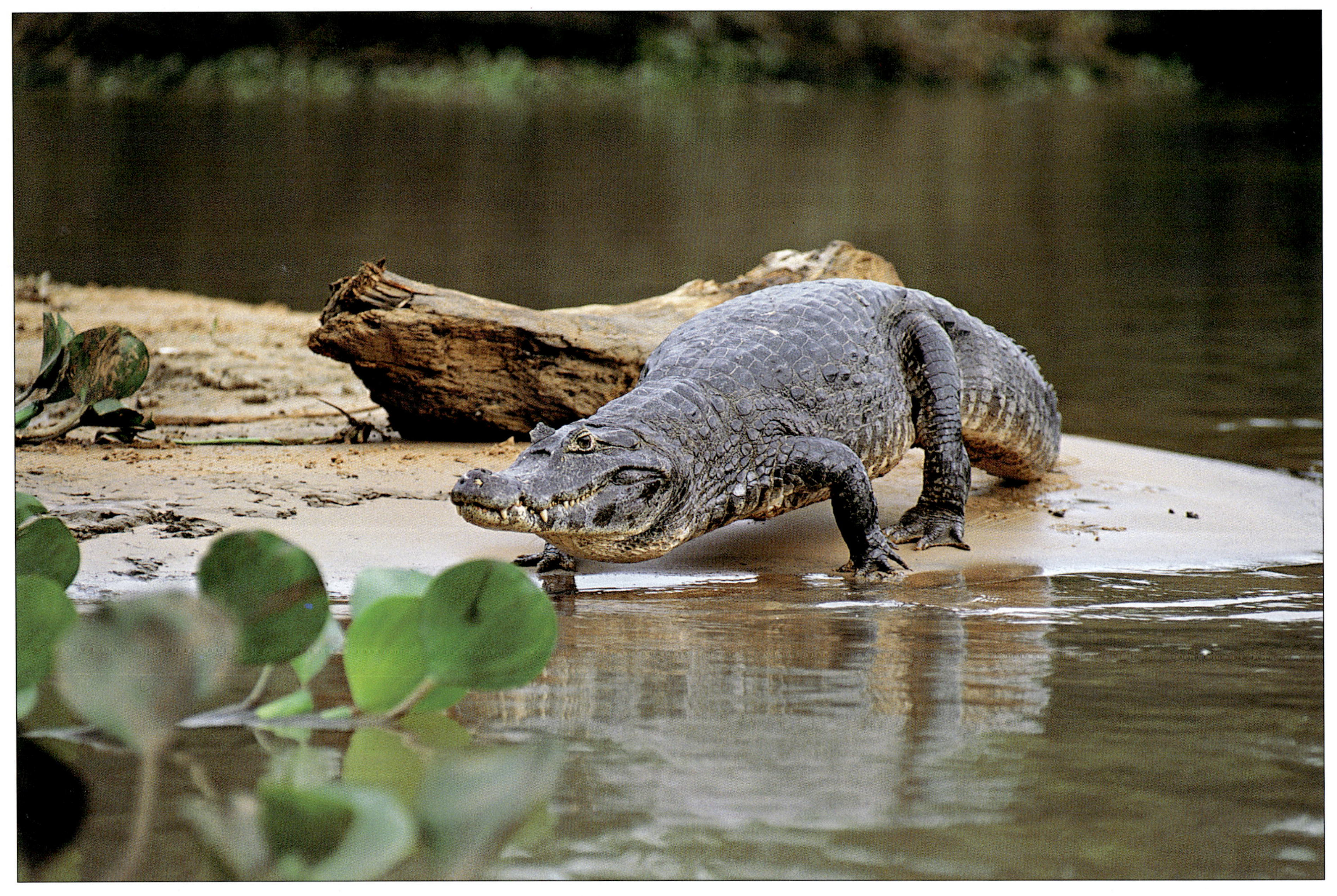

Mature adult Caiman along the river

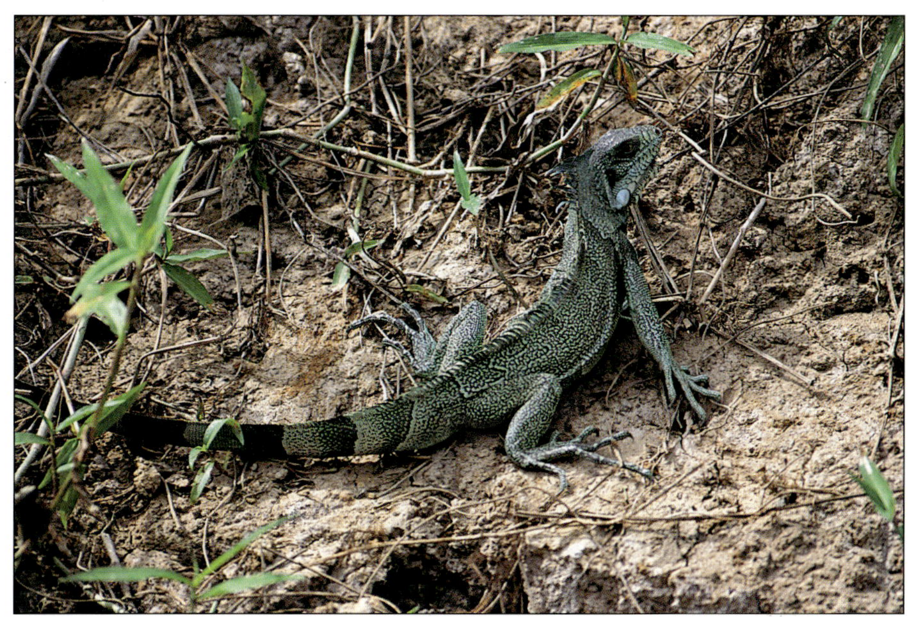

Iguana

Iguana iguana

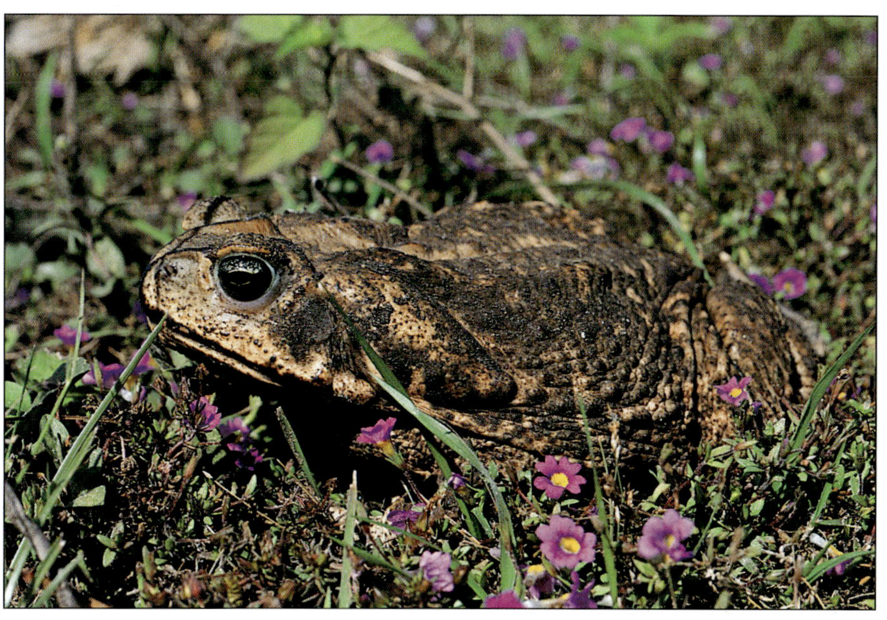

Cain Toad

Bufo marinusSpecies

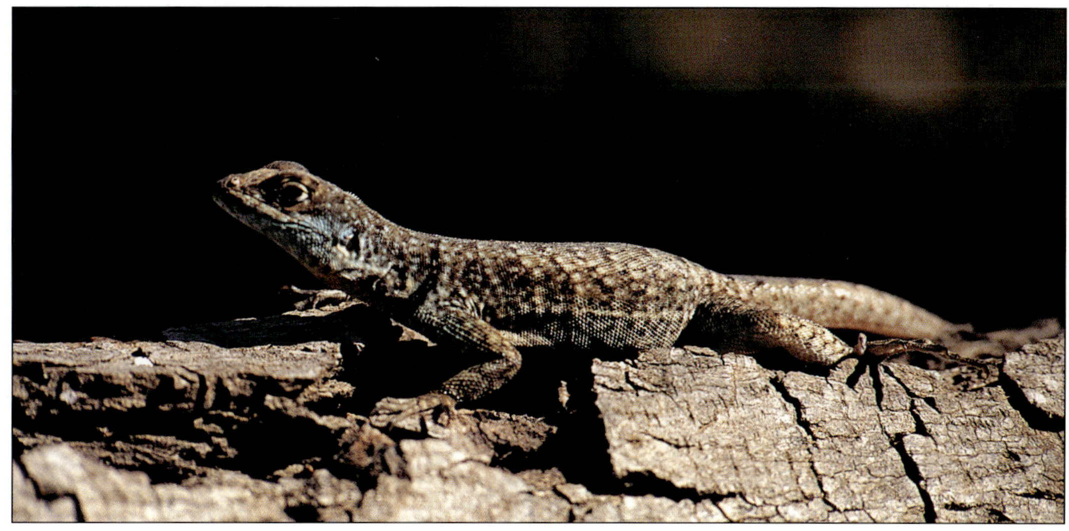

Lizard

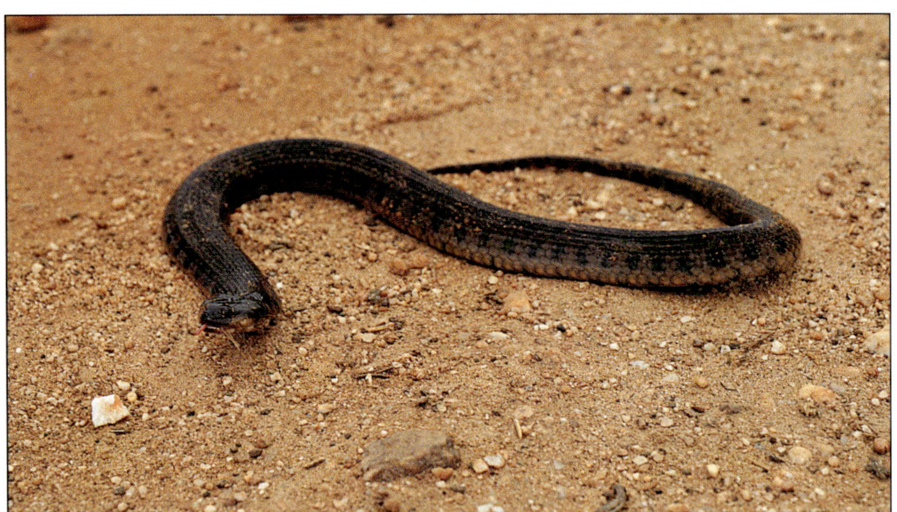

Most snakes are viewed on the road sunning themselves like this unidentified species

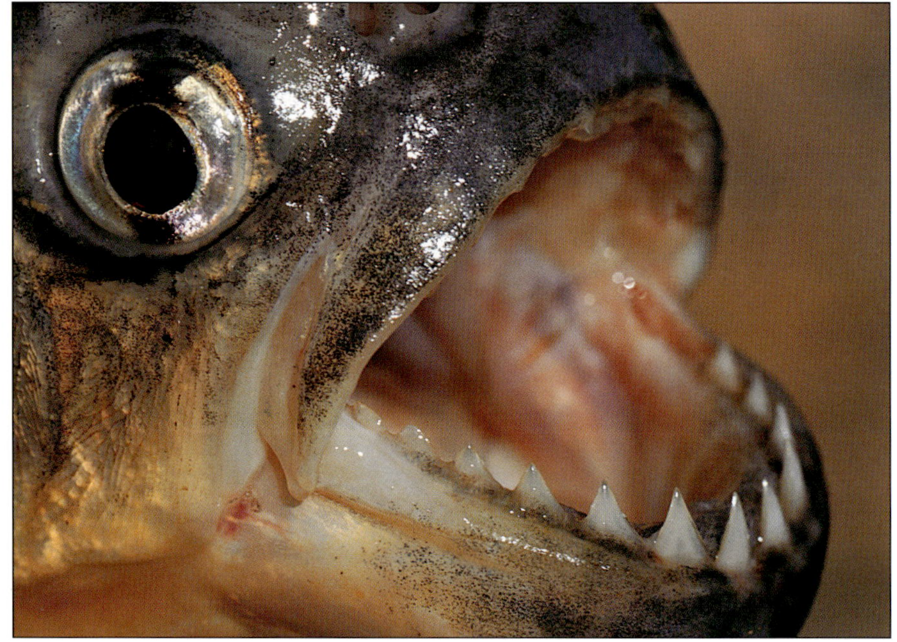

The clouded, slow-moving water in the Pantanal is prime habitat for piranha. Hanging out in small, loosely organized schools of various sizes, they are not as voracious as most people think. Local children swim in the rivers all the time. Their only precaution is not having open wounds anywhere on their bodies. Piranhas are attracted to blood. A smallest trace of blood in the water could start a "feeding frenzy." These fish are most dangerous in the dry season when they are confined to ever decreasing bodies of water. Food becomes scarce and under this pressure cannibalism takes place.

The piranha's teeth are triangular and razor sharp. Teeth of the lower jaw fit perfectly into spaces between the upper teeth, creating a tremendous vice-like bite. The only purpose of these teeth is cutting and biting. They are not used for chewing. Food is swallowed whole, allowing the piranha to eat very, quickly.

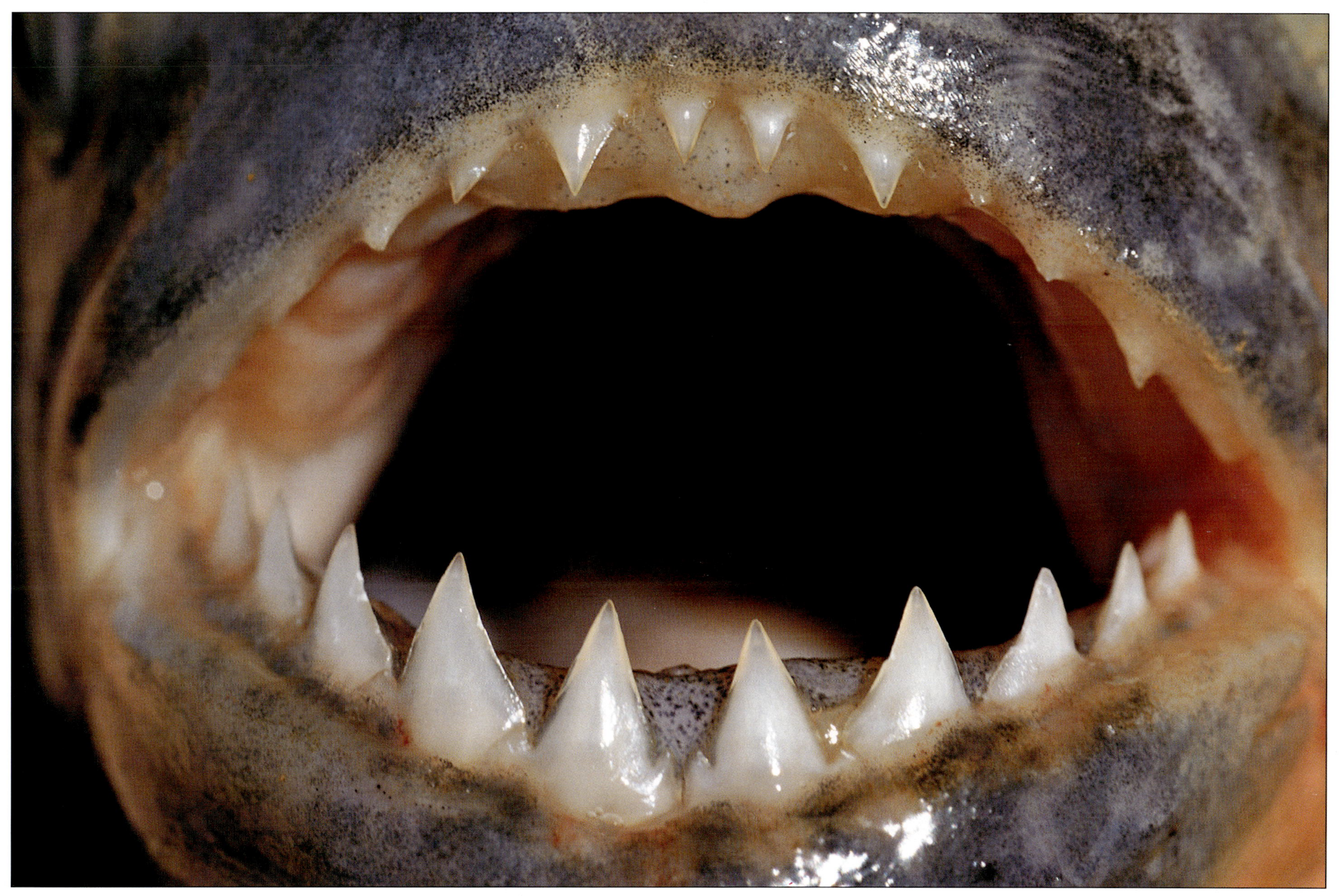

White Piranha
Serrasalmus brandti

Index